SpringerBriefs in Applied Sciences and Technology

T0183090

More information about this series at http://www.springer.com/series/8884

Alexandra Hyard

Non-technological Innovations for Sustainable Transport

Four Transport Case Studies

 Springer

Alexandra Hyard
University of Sciences and Technologies
of Lille
Villeneuve d'Ascq cedex
France

ISSN 2191-530X   ISSN 2191-5318   (electronic)
ISBN 978-3-319-09790-9   ISBN 978-3-319-09791-6   (eBook)
DOI 10.1007/978-3-319-09791-6

Library of Congress Control Number: 2014946198

Springer Cham Heidelberg New York Dordrecht London

Printed on acid-free paper

Springer is part of Springer Science+Business Media (www.springer.com)

Contents

Chapter 1
Introduction

Alexandra Hyard

Abstract The aim of this chapter is to show the main issues of non-technological innovations for sustainable transport and to outline the following chapters of this Brief.

Keywords Innovation · Transport · Environment

Managing greenhouse gas (GHG) emissions from transport is a priority. The transport sector has the second highest GHGs in the EU after energy. According to the European Commission [3], an increase of 74 % is projected for GHG emissions from EU transport between 1990 and 2050. If we examine emissions of CO_2 by transport mode, we could note that emissions from the road sector dominate and are projected to continue to dominate: in 2007, more than two thirds of transport-related GHGs were from road transport. But emissions in the aviation and maritime sectors are growing particularly quickly.

To curb the expected growth in these emissions, transport policies promote innovation but, generally, only technological innovation. For example, the last International Transport Forum (2010), which brought together Ministers, leading decision-makers and thinkers, emphasised technological improvements as the core of climate change policy in the transport sector. The technological innovations that improve fuel economy and transform the energy basis of transport are essential for GHG abatement. But these innovations must not obscure the role of non-technological innovations in reducing emissions. For example, innovations in traffic management or "green logistics" [1] are non-technological innovations that could reduce emissions related to road transport. Compared to technological innovations, non-technological innovations are less visible. Perhaps this is why politicians prefer technological innovations rather than non-technological innovations. But the latter

A. Hyard (✉)
Faculty of Economics and Social Sciences, CLERSE (UMR CNRS 8019),
University of Lille 1, 59655 Villeneuve d'Ascq, France
e-mail: alexandra.hyard@univ-lille1.fr

© The Author(s) 2014   1
A. Hyard, *Non-technological Innovations for Sustainable Transport*,
SpringerBriefs in Applied Sciences and Technology, DOI 10.1007/978-3-319-09791-6_1

also contribute towards the abatement of environmental problems caused by transport. Moreover, non-technological innovations are generally less expensive than the others. Thus, for example, RAND Europe [5] shows that non-technical innovations may contribute cost-effectively to reducing transport emissions. In consequence, non-technological innovations in transport must be taken seriously.

The purpose of this Brief is to focus on non-technological innovations in transport. According to the third edition of the *Oslo Manual*, an innovation "is the implementation of a new or significantly improved product (good or service), or process, a new marketing method, or a new organisational method in business practices, workplace organisation or external relations" [4: 46]. This broad definition of an innovation includes, in fact, four types of innovations: product innovation, process innovation, marketing innovation and organisational innovation. Product innovation and process innovation are closely related to technological innovations. Non-technological innovations refer to marketing innovation and organisational innovation, which are defined as follows:

> A marketing innovation is the implementation of a new marketing method involving significant changes in product design or packaging, product placement, product promotion or pricing.
> An organisational innovation is the implementation of a new organizational method in the firm's business practices, workplace organisation or external relations [4: 49, 51].

As we shall see below, organisational innovation is recommended for sustainable transport, i.e. an economically profitable transport that is also environmentally and socially friendly. These three elements of transport are indeed taken from the three dimensions of sustainable development given by the Brundtland Commission in its report entitled *Our Common Future* (1987). In the report, the Commission defines sustainable development as a "development that meets the needs of the present without compromising the ability of future generations to meet their own needs" [6]. But tensions between the economic, environmental and social dimensions should not be ignored (see [2]). In relation to transport, one environmentally damaging mode of transport might be preferred to another if it is relatively inexpensive. Therefore the challenge of non-technological innovations is also to reconcile these contradictory dimensions.

In order to provide a concise review of non-technological innovations for sustainable transport, the Brief will be based on the main transport modes (road, rail and maritime). Each mode shall be examined through one case study.

Before examining each transport mode the Brief will, first of all, raise the question of the best public policy for sustainable transport. This question is asked by Hakim Hammadou and Claire Papaix in their chapter entitled "Which public policy to move towards low carbon mobility?". The authors show that implementing purely regulatory tools or modifying the generalised cost of car trips through congestion charging, parking charges or public transit fares all influence a modal split in a different way—by the fixed or variable nature (time, area-dependent, etc.) of the scheme considered, the volume of those making journeys, and different time-periods. In addition, the tools presented have different distributive effects on social welfare, some being known for being more regressive than others. At least when acceptability issues are considered, psychological factors dominate self-interest variables. This makes the development of the tool and determining the branded arguments the major factor for accepting a policy measure, and thus for its successful implementation.

Chapter 3 focuses on road transport. Eleonora Morganti and Laetitia Dablanc study recent innovation in last mile deliveries. In cities and metropolitan areas, last mile deliveries are a key factor which contribute to local economic vitality, the quality of urban life and the attractiveness of urban communities. However, the freight transport sector is responsible for considerable negative impacts, mainly with regard to congestion, CO_2 emissions and air and noise pollution. In order to improve efficiency and reduce adverse impacts, city planners and policy makers have launched major initiatives to enhance freight logistics systems, finance organisational and technological innovation, and introduce new traffic regulations. In this chapter, they provide an overview on innovative measures for last mile operations. The authors discuss innovation for delivery vehicles, dealing with new concepts, sizes and technologies (electric and hybrid powered engines). Then they detail recent innovation on parcel delivery services for e-commerce, focusing on pickup points and lockers. Finally they present organisational initiatives on urban food logistics and the deployment of food hubs as urban distribution centres for perishable products.

The next chapter is devoted to rail transport. The authors, Corinne Blanquart and Thomas Zeroual, examine innovative rail services for green supply chains. For them, technological impact fostering sustainability is over-estimated. Transport policies should identify shippers' needs much more. These needs are differentiated i.e. constraints related to micro-economic time and cost optimisation. According to the authors, they involve much more than technological solutions in order to foster sustainable transportation.

In the last chapter focusing on maritime transport, Emeric Lendjel and Marianne Fischman are interested in maritime ports and inland interconnections. Recent research on maritime port hinterlands points out the relevance of mass ground transport modes such as barge transport for the enormous flow of containers to and from harbours, especially when a maritime port is located at the mouth of a river. However, the modal share of container barge transport in French maritime ports (9 % of TEU in Le Havre and 5 % in Marseille in 2007) is significantly lower than elsewhere (32 % in Rotterdam and 33 % in Antwerp). Some reports and studies explain the viscosity of container barge transport flows as a result of several factors, generally concentrated around the seaport community. Continuing previous seminal works, this paper adopts a neo-institutional approach [7, 8] of container barge transport to understand how the factors generating this viscosity are managed. Section 5.2 describes the characteristics of the transaction of container barge transport. Section 5.3 is devoted to its attributes (asset specificity, frequency, uncertainty). According to Williamson's [8] remediableness criterion, the observed governance structure of a given transaction is presumed efficient and aligned to its attributes. Thus Sect. 5.4 deals with observed governance structures of container barge transport chains with a focus on Le Havre, the main French container seaport, and shows how agents try to limit opportunism in ex-post haggling over quasi-rents or under-investments. Implementation of a new institutional environment to modify governance structures is analysed, and a comparison with currently implemented governance structures observed in the Rhine is made. Finally, Sect. 4.5 suggests ways of dealing with the remaining coordination problems impeding the development of container barge transport in France.

References

1. Button K (2010) Transport Economics, 3rd edn. Elgar, Cheltenham
2. Dovers SR, Handmer JW (1993) Contradictions in Sustainability. Environ Conserv 20:217–222. doi:10.1017/S0376892900022992
3. European Commission (2011) EU Transport GHG: routes to 2050. http://www.eutransportghg2050.eu
4. OECD and Eurostat (2005) Oslo manual—proposed guidelines for collecting and interpreting technological innovation data, 3rd edn. Paris
5. RAND Europe (2003) SUMMA, Deliverable 9 of workpackage 2: marginal costs of abatement for environmental problems caused by transport. http://www.tmleuven.be/project/summa/home.htm
6. WCED (World Commission on Environment and Development) (1987) Our common future. Oxford University Press, Oxford
7. Williamson O (1985) The economic institutions of capitalism. The Free Press, New York
8. Williamson O (1996) The mechanisms of governance. Oxford University Press, Oxford

Chapter 2
Which Policy Tools to Move Towards Low Carbon Mobility?

Hakim Hammadou and Claire Papaix

Abstract Most of the challenges associated to the transition towards low carbon mobility being concentrated in cities, this chapter focuses on the implementation of policy tools at the urban scale. After a conceptual overview of the economics of low carbon mobility in Sect. 2.1, we present the toolbox of the policymaker for reducing CO_2 from urban mobility in Sect. 2.2, by subsequently appraising the efficiency, equity and acceptability of a sample of policy tools.

Keywords Transport policy · Low carbon mobility

2.1 Introduction

The uncertainty which weigh on the spatial damages from climate change (the locations of the impacts are not necessarily the same as from where CO_2 emissions are generated), on the time horizon (the next generation might be more affected than the present one) and on the magnitude of the events, makes the CO_2 externality rather difficult to evaluate. Currently, CO_2 emissions account for the relatively lowest external cost from road transportation. For instance, in French transport investments analyses, the CO_2 cost in dense urban areas is estimated at 0.45 c€/passenger km whereas congestion accounts for 16.6 c€/passenger km [13].

However, transport activities represent more than a third of overall CO_2 emissions in the EU-27 in 2009, with an increasing trend since 1990 [27]. Therefore,

H. Hammadou (✉)
EQUIPPE, Faculty of Economics and Social Sciences,
University of Lille 1, 59666 Villeneuve-d'Ascq Cedex, France
e-mail: hakim.hammadou@univ-lille1.fr

C. Papaix
DEST-IFSTTAR, 14-20 Boulevard Newton, Cité Descartes,
77447 Marne la Vallée, Cedex 2, France
e-mail: claire.papaix@ifsttar.fr

© The Author(s) 2014
A. Hyard, *Non-technological Innovations for Sustainable Transport*,
SpringerBriefs in Applied Sciences and Technology, DOI 10.1007/978-3-319-09791-6_2

Europe has established far-reaching ambitions for reducing the risk for climate change and has identified a potential CO_2 abatement field of 60 % within transport activities [24]. The road mobility accounts for less than 80 % of the total CO_2 emissions from transport in 2009 [25]. Furthermore, because most of the distances travelled are made "locally", and this is a growing issue in line with global demographic and urbanisation trends along with climate change effects [17], urban road mobility constitutes the biggest chunk for cutting CO_2 emissions in transport. Indeed, even if kilometres travelled grow more over long-distance trips, just as CO_2 emissions related levels, the bulk of the trips is made in cities and presents the special feature to create other external costs (local air pollution, congestion, safety, noise, etc.) for the society. At the global scale, urban transport accounts for 10–30 % of the CO_2 emissions of cities depending on the level of travel demand, transport supply, technologies, urban form, economic structure, industrial output, and other characteristics of each city [65]. In developing cities, characterized with high transport demand and an overreliance on inefficient transport systems, this share can be as high as 50 % (as e.g. in Mexico City; whereas in Beijing and Shanghai, carbon emissions from transport represent less that 10 % and other pollutants are more predominant). In developed cities, with saturating travel demand and more performant transport systems, the share can still be around 20 % (e.g. in London and New York City).

Most of the challenges associated to the transition towards low carbon mobility being concentrated in cities, this chapter focuses on the implementation of policy tools at the urban scale. After a conceptual overview of the economics of low carbon mobility in Sect. 2.1, we present the toolbox of the policymaker for reducing CO_2 from urban mobility in Sect. 2.2, by subsequently appraising the efficiency, equity and acceptability of a sample of policy tools.

2.2 Background on the Economic Principles of Low Carbon Mobility

2.2.1 Transport Investments and Financing Issues

In the past, road investments have been dominating, leading to a non-sustainable development of transportation systems. The resulting induced demand has then further encouraged such a non-sustainable pathway of investment decisions in transport.

Road infrastructures account in 2008 for the highest share of annual transport expenditures (see in Fig. 2.1), even if this has decreased over the last decade (from 62 % in 1995 to 58 % in 2008; [26]).

The main source of financing for transport infrastructures investments has traditionally been the State. However, with the current scarcity of public funds, resorting to private actors has become a necessity. In Sweden for instance, 67 % of the road infrastructures financing comes from the private sector [39]. In general, Public-Private Partnerships (PPPs) take the form of a public service delegation

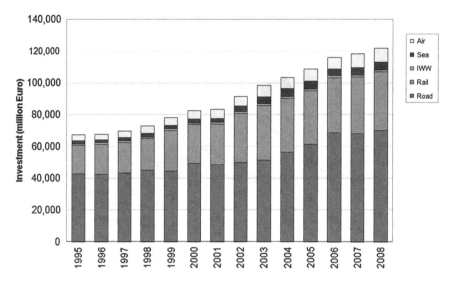

Fig. 2.1 Investments in transport infrastructure (millions of Euros) in EEA member countries. *Source* EEA [26]

contract over a long period, as will be further discussed in section "Public and Private Actors". On the presentation of the public and private stakeholders of the low carbon mobility system.

2.2.2 The Relation Between Economic Growth, Transport Activity and Carbon Emissions

Low-carbon mobility refers to a lesser carbon intensive mobility [33]. This can be achieved by the use of three means (with the different effects shown in Fig. 2.2):

- Changing the social norms to move towards a lower level of mobility (the most difficult pathway).

The key to this pathway is a fundamental shift in societal values that would result in alternative forms of production and consumption, and perhaps also in different forms of car ownership (shared rather than individually owned). This change would lead to a decline of the consumerism and to a global reconfiguration of the production chains. As the 'lifestyle' structurally changes under this scenario, there is no need any more for decoupling the mobility from economic growth (that now follows a different pattern) or from carbon missions (already disconnected from economic growth). Such changes in societal norms give in particular a larger priority to "slower travel", and trigger new forms of economic activity: "slower is better" and "close is better".

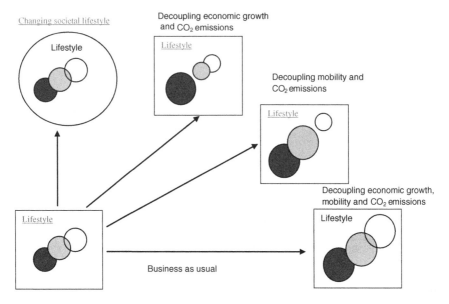

Fig. 2.2 The three spheres of economic growth, mobility and CO_2 emissions. *Source* Givoni and Banister [33]

- Following an economic growth path that is less dependent on transport activity.

Economic growth and transport activity are closely related. This means that a higher volume of transport accelerates economic growth; and respectively, that a rise in transport activity stems from economic growth. However, there is a need for keeping or increasing the economic growth rate without increasing mobility levels to the same extent. To do so, it is possible to preserve the "current lifestyle" but in a way that assumes small changes in daily life not to unnecessarily increase transportation needs. Moving from the standard "globalization" model to a "globalization" model (local/regional autonomy of production; where the distribution and consumption of goods and services relies on shorter trip distances) could provide an answer to this challenge. The key of this trajectory is the shortening of the trip distances. Therefore, home-based work, teleworking, or video-conferences could be strongly recommended.

- Transforming the transportation system to produce less carbon emissions.

To attain low carbon mobility is to decarbonize transport via technological developments, by example the Zero Emission Vehicles. This pathway is attractive as it can be an important driver for increase economic activity, through investments in green mobility technology. The low carbon mobility challenge assumes that environmental and economic objectives do not necessarily contradict each other, and can be attain in the same time. This pathway does not require systematic change but much more research and development (R&D) efforts in surface transport, and car in particular, are focused on efficiency and not on speed, carrying capacity (size) and energy-consumption. The public sector also has an important role to play in supporting and incentivizing (through pricing, subsidies and other economic tools) the private sector and by adopting policies, like widespread adoption of Low Emission Zones (LEZs).

2.3 Efficiency, Equity and Acceptability Criterion of a Sample of Policy Tools

2.3.1 Different Ways of Classifying and Selecting Instruments

2.3.1.1 Depending on the Decision-Makers Involved, Policy Targets and Structural Designs of the Instruments

Public and Private Actors

Low carbon policymaking can emanate from both from public and private actors. If we essentially focus in this chapter on the public policy tools, it is worth mentioning the role and strategies of the private actors too.

Public Private Partnerships (PPPs) illustrate the case of a structured financing source that allows to alleviate the public budget by transferring the risk towards private partners. PPPs have increasingly developed in the past few decades in some European countries, particularly in the UK and Portugal (forerunners of PPPs; [63]), Italy (with the introduction of tolls on major roads leading to central cities already in the early 1920s) and France (where over 70 % of the urban public transport networks in 2011 are operated using a public service delegation).

Open competitive tendering processes can also constitute a public-private solution for rolling out low carbon mobility. In Stockholm County, the use of biofuels in regional buses is supported by legal contracts between the bus operator and the tendering agency. Because the proposed bids are evaluated in the light of their 'environmental properties', bus providers are incentivized to invest in low carbon fuels in order to benefit afterwards from advantageous conditions for operating their fleet [7, 30].

Private actors can also unilaterally develop low carbon strategies. As an illustration, Paris has developed a car-sharing service named 'Autolib'. Autolib is also the name of the mixed syndicate who signed with Bolloré Group a public service delegation that aims at offering an "ecofriendly" transport service, proposing an alternative transport solution, decreasing the use of the private cars and making this service accessible for everybody (see [50]).

Another range of strategies often developed by private companies are the Corporate Mobility Plans (PDE), offering a great potential to deter employees from driving alone and thus to reduce the related CO_2 emissions.

Supply-Side or Demand-Side Targeting

Policy levers can then play either on the demand-side and/or on the supply-side of the low-carbon mobility system's stakeholders. As further detailed in the report of Meurisse and Papaix [50], instruments can be classified according to their targeted groups:

1. Tools such as speed limit measures, Low Emissions Zones (LEZs), High Occupancy Vehicles lanes (HOV), parking access management, pricing schemes related to vehicle purchase, ownership or use, fuel pricing, road user charging, parking fees, energy consumption and CO_2 emissions labeling for new passenger cars, etc. apply to road users—regarding the demand-side;
2. Tools such as CO_2 emissions standards, obligation of a minimum content of biofuels in fuels, car tyre labeling, etc. apply to industrial actors; and instruments e.g. the binding information to report on CO_2 emissions from transport services, eco-driving training, etc. apply to transport professionals—regarding the supply-side.
3. A third category of policy levers could be the one applying to local transport authorities (see e.g. the norms on publicly accessible charging infrastructures for electric vehicles).

'Hard' Versus 'Soft' Measures

Pricing signals may, at least in the short term, have a limited effect on travel behaviors especially on the choice of route, destination, modes and trip frequency regarding the use of congestion tolls; and vehicle ownership and departure time in the case of parking charges [64]. Therefore in some cases, the so-called 'soft measures' such as information, education, marketing and communication policies can be used for redirecting user practices more efficiently than when using 'hard measures' (e.g. subsidies for technological innovation, transport infrastructure investments, etc. see [5]). Besides, the study of Xenias and Whitmarsh [66] on the preferences of political experts and civil society regarding the use of instruments for low carbon mobility reveals that qualitative and demand-side management tools were better accepted than technical-and-economic levers.

The individualized marketing operation set up by the Region Picardie in 2012 [68] confirms this argument and adds the 'trial' phase to the policy recommendations. Following from the experimentation carried out in September 2012 and according to the two-waves survey before (from July 2012) and after (from October 2012) the operation, 60 % of the total surveyed participants continued to use the train after the end of the pilot experimentation, 40 % shifted from car use to other modes, 37 persons over 150 purchased a commuting/leisure subscription and 18 over 150 bought single tickets (because they didn't find appropriate subscription cards).

First-Best and Second-Best Features

'First-best' and 'second-best' adjectives can alternatively denote the economically optimal *a*—conditions for policy implementation and target definition; or *b*—the policy-tool design itself.

On the first sense, establishing a policy "*given that all other parts of the economic system are working perfectly and distributional matters are not a contentious issue*" [10] can be a good definition of what is a first-best ("academic ideal") environment. Where and when major legal or social imperfections come into play (generally following the introduction of the scheme in practice), one can prefer second-best arguments for policy recommendations. Second-best policies generally aim at correcting such full short term and long term effects, as markets' multiple overlaps (e.g. between transport, urbanism and labor supply), external effects interactions (e.g. between CO_2, local pollutants and road safety, etc.), etc. De facto, their underlying instruments do not pursue one objective only (e.g. reduce CO_2) but interfere with several at once (reduce or increase environmental externalities, accessibility, territorial attractiveness, etc.). Since at the scale of urban mobility, suboptimal equilibrium conditions are frequent, we choose to focus on *second-best* policies (i.e. on instruments that can play positively or negatively on several goals at once) instead of *first-best* ones (one tool per policy target).

On the latter sense, policies can be qualified of second-best not because distortions exist elsewhere in the spatial economy but by the 'design' of their schemes. Besides, note that their net welfare effect can actually be better than the one of first-best schemes—this is the example of flat cordon pricing schemes in mono-centric cities over distance, route or time differentiated tolls [62].

2.3.1.2 Efficiency, Equity and Acceptability Properties of the Tools

Conditions for Their Economic Efficiency

When looking at the different sequences of the overall travel demand choices' formation process, (alluding to the traditional *Four-stage model* of Ortuzar and Willumsen [52]), 'trip origin choices', 'trip destination choices' (the two latter affecting 'vehicle ownership choices' but also 'land uses' on the longer run), 'travel mode choices' and 'route choices' (also potentially changing e.g. 'departure time choices' on the shorter term) are differently impacted by the policy-levers. The travel mode choice step appears to have the largest room for policy-action, i.e. to be the easiest way to convey price signals judging from the number of suitable policy-tools [56, 64].

In addition, if we refer this time to Schipper et al. [58], shift from car use to low-emitting modes is the term of the Activity–Structure–Intensity–Fuel (*ASIF*) equation for CO_2 emissions mitigation in transportation that occurs the earliest in the time-horizon (Structure), compared to vehicle efficiency improvements (Intensity), switch to biofuels (Fuel) or lesser travel activity (Activity), and that brings into play the lowest degree of industrial actors and strategies and thus of market conditions requirements.

If all the different steps are constantly in dynamic interactions, we focus in this chapter on the mode choice step as a central policy target in order to appraise the efficiency of policy tools. However, one can note that other means can participate as effectively to the reduction of CO_2 emissions such as car-sharing, by increasing of the loading factor within a same mode [31, 46].

Echoing to what has been said before on the multi-leveled externalities from urban mobility, the economic efficiency of a tool can either reinforce the global performance of policymaking or run against the implementation of another instrument. On the former case, May et al. [49] identify four ways in which policies (parking charging, congestion pricing and additional measures) can positively interact with each other:

• Complementarity: the use of two instruments has greater impacts than the use of either alone;
• Additivity: the benefit from the use of two or more instruments is equal to the sum of the benefits of using each in isolation;
• Synergy: the simultaneous use of two or more instruments yields higher benefits than the sum of the benefits of using either one of them alone (Additivity and synergy can be considered as two special cases of complementarity [49]);
• Substitutability: the use of one instrument completely eliminates any benefits from using another instrument.

The aforementioned interacting effects between the instruments worth being considered in the appraisal of the tools. In this respect, the accompanying of 'pull' measures (disincentives) by 'push' tools (incentives) is known for reinforcing the success of policy implementation [40].

Equity Effects of the Policy Levers

No consensus has been found in the literature for defining the equity concept in the transport sector [4]. The equity goal itself is often vague and this is usually due to the difficulty of implementing it or measuring it in practice [48]. In France, social and territorial equity are specified in reference texts (for example in [14]). One of the proposed indicators for appraising it is the ratio between the total surplus created by a transport project in the zone (e.g. time gains, pollutions cost savings, etc.) and the total income of users in this particular zone of impact. To deal with fairness, a special attention is recommended to be paid to the initial rent situation of individuals, for instance when the transport network is structurally servicing more certain categories of individuals than others. Then, redistributive effects from pricing policies in particular are also worth being considered since transport costs can already be significant in households' budgets (especially for the less well-off). Transport fuel expenditures represent for instance 12.7 % of the most vulnerable households' annual income in France in 2006 [37]. Hence, the preservation of individuals' capacity for lifestyle adaptation and reliance to available transport alternatives following the introduction of an economic tool constitutes a good focal point for assessing the equity of the scheme at focus. Regulators should be careful to avoid adding a damaging "carbon bill" on these low-income households, who are most of the time and by a majority structurally price-inelastic and highly car-dependent [12].

Following the approach of Martens [47], three questions can be used for appraising equity effects from policy implementation: (1) Which goods and bads or benefits and costs should be at the focus of the equity analysis?; (2) How should 'members of society' be conceptualized, i.e. which population groups should be distinguished?; and (3) What constitutes a 'morally proper distribution', i.e. which yardstick or distributive principle should be used to determine whether a particular distribution is fair?.

Acceptability Challenges When Implementing the Tools

The attitude construct of *acceptability* designs "support, agreement, feasibility, to vote for, favorable reaction" to a particular scheme and "describes the prospective judgment of measures to be introduced in the future". By comparison, public *acceptance* refers to the "behavioral reactions [of respondents] after the introduction of a measure". Moreover, the adjective *public* can involve, depending on the studies, "motorists, voters in general, consumers, citizens or inhabitants" [57].

Looking at the literature on the theory of planned behavior [2], the key findings are that socio-economic characteristics of individuals and transport network related variables explain only a little part of the acceptability of congestion pricing [57]. Attitudes factors remain the largest predictors for policy acceptability. Specifically, the gap between pro-environmental attitudes and acts, i.e. a positive reaction towards policy levers aiming at reducing e.g. environmental externalities, can be explained by five main variables:

1. The absolute importance of the issue. This depends on 'personal norms' (i.e. the general pro-environmental orientations of individuals as measured by e.g. the New Ecological Paradigm scale; see [23, 59]), especially for the push measures; followed by the 'social pressure' effect (the fact that most people strive for social integration, conformity and consonance; see [29] making them more willing to accept the unavoidable), and by other wider preferences, towards risk and uncertainty (see [6]) for example.
2. Environmental outcome desires [from the policy]. Specific policy beliefs depend on (i) the knowledge about options of the scheme (user awareness is loosely related to user acceptance; see [57]); (ii) the perceived effectiveness and efficiency of the proposed measures (respectively potentially influenced by experience/familiarity; the labeling and defined objectives of the scheme [41]; the use of the tax-revenues [56], or territory coverage); (iii) and its perceived

fairness.[1] Tax resistance (iv) can also explain singular attitudes regarding to the previous beliefs (see e.g. [57]).

3. Self-efficacy in solving environmental problems. This depicts the self-reported role of the individual according to his subjective representation of the responsibility sharing to protect the environment [57].

Generally speaking, public acceptability for *pull* measures (policy-instruments aiming at reinforcing the attractiveness of alternative travel options) tends to be higher than for *push* measures (continued opposition to coercive tools aiming at deterring car use; see [23]).

The different ways of classifying policy levers having been presented, along with the criterion for choosing among them, we test this framework in what follows on a sample of instruments. Low Emissions Zones, congestion tolling, parking charging and public transit faring schemes are selected and their economic, equity and acceptability evaluation is, whenever possible, supported by evidences from real experiments.

2.3.2 Sample of Instruments

2.3.2.1 Low-Emissions Zones

Economic Efficiency

A tremendous variety exists among the two hundreds Low Emission Zones (LEZ) operating in Europe [15, 16] in terms of perimeter size, nature of the restriction (total restriction or access fee), polluting vehicles targeted (usually depending on the Euro class, weight or age of the vehicle), and temporality (all day or business/peak hours only). LEZ specifically target local air pollutions e.g. particle matters and nitrogen oxide emissions (also noise from 2002 in the case of Stockholm; [22]) but can have indirect impact on CO_2 as well, by modal shift effect. Referring to the classification of instruments above, the LEZ pertain to the command-and-control levers playing on the demand-side.

If we focus on the London and Stockholm case—both combining an interdiction of polluting trucks over the greater metropolitan area (with increasing standards over time) and a urban congestion charge—the following air quality impacts, socioeconomic costs and car fleet changes can be highlighted.

In Stockholm, the restricted areas apply nationally to all trucks of more than 6 years-old (except those between 6 and 8 years-old if they belong at least to the Euro 3 class) and to buses of more than 3.5 tons since 1996 [1]. Four years later,

---

[1] Equity perception is obviously differently perceived depending on the 'evaluator'. It will for example vary if it comes from frequent car users by themselves and for themselves or by themselves and for low income groups or citizens in sparsely populated areas.

PM10 emissions were reduced by 40 % (their concentration by 3 %), and the emissions of NO_2 by 10 % (their concentration by 1.3 %). Looking at the vehicle fleet composition, an energy substitution has occurred in Stockholm following from the measure, with an observed drop in gasoline fuels to the favor of more diesel and LPG fuelled trucks and a drop in gasoline and diesel to the favor of ethanol and LPG buses. The phenomenon has been accelerated since trucks have increased while buses have decreased over the observation period.

Results were more nuanced in London (implementation of the scheme in 2008), where the LEZ only reduced by more than 30 % (according to modeled results) the area of London that was exceeding the annual threshold of PM10 regulatory concentration. The reduction was lower for NO_2, even though the concentration is more problematic than PM10 in London, and didn't improve the local air quality in general. Conclusions were more positive when a more stringent policy was simulated (inclusion of taxis and buses in the measure). The fleet turnover for affected vehicle classes in London increased substantially when the zone was first introduced before returning to the national average in subsequent years [22]. Despite an overall growth in freight vehicles operating in London, the number of pre-Euro III vehicles has dropped and this has been coupled with a switch from rigid vehicles to light commercial vehicles and articulated vehicles.

Regarding financial issues, implementation costs in Stockholm were twice cheaper than expected (the return on investment ratio being the highest when the share of trucks is large), and 80 % of the operating costs were compensated by the environmental gains. In London, costs were about five times higher than in Stockholm (in absolute terms, without relating it to the size of the area to regulate). Infraction fines (and charge revenues in the case of London) are not, in neither of the case, subjected to specific public expenditures outlay [1].

Equity Effects

City-wide restriction schemes tend to cause less spatial unfairness than traffic exclusion from city centres only [51], by setting all the polluting vehicles on an equal regulation and avoiding giving more adaptation capacity to some actors than to some others.

Acceptability Challenges

Again, territorial and social equity largely drive the acceptability of the scheme [19]. But before considering the trip-maker's perspective, the strength of the ongoing health-based air quality standard requirement in the country principally motivates the interest of the policy-makers in implementing the LEZ (tightening of the policy context recognition in Europe (see [51])). Then, the seriousness of the air pollution evidences in cities, transparency, understanding of the measure and exemplarity favor the public acceptability of LEZ, when it is not too much hampered by commercial trucks lobby pressure (high costs on the businesses). In this regards, since freight haulers in London were the principal target of the charging scheme, they expressed their disapproval.

2.3.2.2 Congestion Tolling

Economic Efficiency

Urban road charging is a sound instrument usually recommended by economists raise revenue, reduce traffic congestion, ration road space, improve the local environment, mitigate climate change, and enhance social inclusion and equity through the pricing of the social marginal cost of a trip. Tolling vehicle drivers who enter a specified geographical zone for the cost of the congestion they impose on other drivers is indeed a useful instrument to deter from car use and encourage low-carbon emitting modes. Estimates from the Stockholm charging trial introduced in January 2006 show for example [21] that close to one-fourth of the work trips by car passing the cordon disappeared (between September 2004 and March 2006), of which the big majority moved to public transit and the rest adapted to the scheme by changing frequencies, combining trip purposes and increasing trip chaining.

Adding more to this conclusion, the literature review from Li and Hensher [45] on the impact of congestion pricing on travel behaviors essentially shows that changes in departure times was the major effect from the scheme (when it is time-differentiated, for example in Stockholm), followed by reduced car use, modal shift and relocation of work and/or residential activity. Besides, the use of congestion tolling also avoids having to resort to road capacity investments (to address congestion) that usually induce road traffic (the 'Downs-Thomson paradox', see [20]) and therefore other negative externalities (namely environmental impact, unsafety or infrastructure use).

Referring to the conditions of Gunn [34] for a perfect implementation of a policy measure, Ison and Rye [40] highlight the following success factors relating to congestion charging after reviewing real experiments across Europe:

External circumstances—e.g. the quality of the public transport system, revenue perception and use, technological conditions and the severity of congestion—strongly play on the level of public acceptability. As an illustration of the technological factor, the technology used in the electronic road pricing scheme (ERP) in Hong Kong raised public opposition due to the fact that it was a western European (British) patented technology [40], thus undermining the trust of the population with regard to the overall introduction of the scheme. On the latter condition, the low level of congestion in Cambridge for instance partly explains the failure in implementing the scheme in 1993. To the contrary, the excessive congestion level in Bergen (relatively to the size of the city) was key in the approval of the scheme. Nevertheless, some authors tend to nuance this idea: Eliasson [21] claims that there was no statistical evidence in the case of Stockholm of a relation between the level of congestion and the degree of accepting the toll.

The availability of financial resources (operating expenditures, administration and enforcement costs—in particular for congestion metering on beforehand of the implementation of the scheme and public transit system strengthening afterwards) along with a **good traffic predictability** (limited by the intrinsic uncertainties due to the dynamic pattern of trip-makers adaptation strategies) generally secure the core functioning of the scheme.

A consistent theory of the cause (need for an analysis of the nature of the problem—what drives demand for private transport and traffic congestion), **and of the effect of the policy** (more visible if the implementing groups—i.e. county, city, district councils, etc.—are cooperating on the measurement of the results and have an interest in metering the outcomes) should be communicated to the individuals.

Objectives of the scheme and use of the revenues should be clearly stated. New elections or political instability to a larger extent (e.g. opinion divergences between the electorate and the politician, opportunism of the decision maker and associated moral hazard and adverse selection problems; see [18]) can affect the goals—and even the existence—of the scheme. Edinburgh, Birmingham, Manchester or New York's attempt cases illustrate such failure. However, changes over time in the definition of the scheme's objectives may happen, in line with the new political agenda, without being detrimental. For instance in Norway, the Oslo, Bergen and Trondheim cordon tolls schemes' objectives have moved from road investments and public transport improvement funding to gridlocks reduction, as a result of growing congestion problems and were still well accepted by the population. Secondly, political uncertainty with respect to the use of revenues can increase the probability that voters will be against the introduction of the toll. To be noted in this regards that public transport subsidization is preferred over toll-revenues redistribution to all voters [18].

Dealing with the enforcement of urban toll and the planning of its objectives, the specification and **correct ordering of the corresponding tasks** is an additional challenge in the case of congestion charging since experiences abroad are relatively poor [even if practices have largely increased over the last years since Singapore (1975), Bergen (1986), Oslo (1990) and Trondheim (1991) with: London (2003), Stockholm (2006), Durham (2002), Milano (2008), Rome (2001) and Valletta (2007); The Netherlands, Copenhagen, Budapest, Gothenburg, Djakarta and San Francisco (to be planned)] and thus offer little possibility of comparison.

Equity Effects

Dealing with equity, urban tolling seems to offer more flexibility for the trip-makers than other schemes applying to all car trips on a same basis (e.g. carbon fuel tax), since individuals that have a lower value of time have the possibility to change their itinerary, their mode or to differ their trips during the day, rather than traveling during the time/over the road/with the mode that is charged. However, travel time savings valuations should be treated with a special care (should be differentiated by 'Social Price of time' groups, as advocated by Galvez and Jara-Diaz [32]) when one wants to implement the congestion charge and to appraise equity effects from user benefits redistribution.

Acceptability Challenges

The acceptability of the congestion tax can be low, particularly in France [55], equity issues being essentially at the core of the refusal. In addition, if toll revenues can be hypothecated to public transport improvements (as for parking fees), **political acceptability** of urban tolling is usually lower than for parking charging [67], especially due to its wider charging coverage (e.g. targeting the commuting staff only in the case of parking measures in the frame of a corporate travel plan for example; see [40] vs. a whole region in the toll case).

De Borger and Proost [18] show that voting patterns in the case of an hypothetical referendum crucially depend on the modal choices of voters—leading to specific expectations with regards to tax-revenues recycling dispositions.

Other authors (e.g. [3, 40, 44], referring to the Trondheim and Bergen cases) raise the issue of public transport provision—and satisfaction—as being all important. Moreover, if it cannot be proved that higher income groups better support pricing strategies, one can observe that the latter groups are more likely to expect advantages from the strategies, whereas lower income groups tend to expect disadvantages [57].

2.3.2.3 Parking Charging

Economic Efficiency

Automobiles tend to be parked for 95 % of the time, either using on-street public parking (charge-free but of limited resource) or private off-street parking [64]. In addition, residential off-street facilities are usually provided in excess by building owners, in line with the high requirements from local housing regulators and their belief that a tight link exist between dwelling choice and level of parking services. As a result, and this is particularly true in areas with low vehicle ownership [60] like city-centers, land use can be inefficiently occupied by barely used parking slots. Thus, parking policies deserve a special attention and can be perceived as a low-hanging fruit for mode shift and CO_2 mitigation—and in particular residential parking.

To add more on the suitability of parking charging, Bonsall and Young [8] emphasize in their literature review that urban parking policies contribute to six goals at the time: "*healthy economic climate; efficient use of transport and land resources; ease of mobility/accessibility; equity of resource distribution; improvement of environmental quality; and enhanced amenity/cultural attractiveness*".

Dealing with road use and congestion challenges (and with *concentrated* congestion in particular, i.e. when much of the traffic is terminating in a same area), because it has less distributional consequences, lower costs of operation and is easier to regulate than urban road pricing [10], parking charging *"may appear preferable as a second best device for containing congestion and other externalities [than urban toll]"*. Furthermore, since 'parking problems and costs' appear to be the number one reason to switch from private car to public transport, followed by the personal car availability and public transport faring and frequency policies [36], increasing travelling costs of automobiles through parking charging is usually recommended by economists [42] to trigger mode shift to mass-transit.

However, such policy goals (e.g. environmental objectives, accessibility, public transit system's performance, or regional attractiveness) can be conflicting. In its political economic analysis, Button [10] highlights the difficulty to sort out the different policy-objectives of parking activity regulation (since the public or private governance can change of policy goals, particularly in dense areas where ever higher space constraints must be combined with accessibility extension for disabled persons or goods delivery issues) and to understand the nature of the other markets involved (i.e. road traffic flows but also local businesses, land use patterns, etc.). Moreover, applying the Gresham's Law to parking management, the author explains the inefficient allocation of road space as a consequence of regulators tendency to distribute parking slots according to the willingness of individuals to spend *time* for parking purposes (the "bad currency", notably due to queuing effect and resulting congestion) rather than (driving out) spending *price* (the "good currency", better reflecting the marginal opportunity cost of the resources involved and clearing the car parks market).

The lack of homogeneity between municipalities' decisions and the regional scale policymaking adds more to the governance challenges. For instance, street parking regulation is generally part of the road system management but it can also belong to the wider transport network regulation or land-use policy of a community. At least, real time information on the availability of spaces and/or on pricing rules can be missing and lead to asymmetrical information problems between off-street parking providers (who possess the information) and parking spaces "consumers".

Beyond governance and institutional challenges, some other factors need to be considered upfront for a successful implementation of parking pricing schemes [8]. Among them:

- *The spatial adverse effects.* A first illustration is the fact that parking activity can be diverted from the charged area onto un-priced nearby streets, where negative external effects from car use can be stronger (air pollution, congestion, noise, phenomenon of "urban heat island", etc.). Second, bad geographical coverage of motorists, e.g. as an effect of subsidized parking spaces at workplaces which puts a significant target of car users out of the pricing control, can hamper the efficiency of the scheme. At least, because the charging regime is not, by definition, distance-differentiated, parking pricing can play over the long run on housing decision (the additional fixed fee rendering longer trips relatively less costly than shorter ones), thus leading to urban sprawl, and potentially re-increasing overall emissions.

- *The hysteresis phenomenon.* Indeed, people's behaviors are difficult to move to optimal ones due to stranded cost and uncertainties. This feature is characteristic of any new policy implementation's success conditions, and in particular for parking charging [10].

Equity Effects

Because parking fees take the form of a fixed amount added to the generalized cost of a trip by car and impact proportionally less individuals making longer trips, equity and acceptability issues can be raised if one assumes that high income groups are commuting longer distances (particularly in American cities; see [8]).

Another unwanted consequence deals with the worsening-off of local economic activity (e.g. the suppression of shopping trips and adverse impact on retail turn-over; see [8]).

Acceptability Challenges

Increasing parking fees, especially at workplaces—as part of a corporate travel plan for instance—is known for leading to high staff opposition, particularly in the public sector [40]. Nevertheless, empirical studies (e.g. the on-worksite car parking charging experiment in the Netherlands; [40]), show that opposition to such measures usually vanishes right after their introduction—as long as they are carefully designed (e.g. income-based charges) and that revenue hypothecation directly or indirectly benefits to the commuting staff (e.g. transit improvements for the journeys to work, etc.).

2.3.2.4 Public Transit Faring Policy

Economic Efficiency

When first best tools to combat congestion are not available (i.e. congestion charges), reforming public transit prices can be recommended to act as a second-best measure [53]. Effects of the measure on public transit patronage are relatively rapid, Bresson et al. [9] showing for example for France and the UK that 99 % of the adjustment is realized within 6 years, especially when transit fares were high.

Additionally, lowering transportation fares through subsidization encourage more economic activity [53]. The resulting decrease in the average price of goods and services compensates the distortionary effects and efficiency losses observed on the labor market from the subsidies, the former effect implying an increase of the real wage and to a higher gratification of the work effort. Consequently, less congested roads lead to higher commercial speeds and a barrier-free public transport more broadly enables fleet operations savings, in terms of e.g. controlling costs, boarding time, etc. [11].

However, there is no unanimity in the literature on whether higher or lower transport prices are better for the social welfare [43]. Indeed, if low prices divert car trips during peak period and allow scale economies (seat occupancy in existing busses) during off peak period, Proost and Van Dender [54] argue that the marginal cost of public funds for subsidizing public transit lead to the largest deficit. Additionally, controlling for (i) changes in supply (needed extension of the public transport network, addition of priority lanes and increase of service frequency), and assuming (ii) that cross-elasticity dominates the direct elasticity (i.e. to bring about an equivalent impact as following from a car deterring measure, a larger public transport fare reduction would be required to increase public transport patronage) and that (iii) elasticity of public transport demand with respect to level of service variables is systematically higher than fare elasticities, Cats et al. [11] nuance the efficiency of the scheme. Using the case of Tallinn, the authors notably found that the measure itself has accounted for only 1.2 % of the public transit ridership increase.

At least, three shortcomings of the scheme can be mentioned:

- A differential fare scheme could have better attracted demand to underutilized segments of the public transport service and avoid supply increase problem in the peak hour where the marginal operational cost is the highest;
- Short-distance trips public transport may become a substitute for walking and cycling rather than car trips;
- A fare-free system can also encourage the population to fraud and to register in the inner city to benefit from the scheme leading to higher operational costs on the long run.

Equity Effects

Farber et al. [28] show for the case of Utah that shifting from the existing flat fare scheme of the PT pricing to an hypothetical distance-based fare structure disproportionately and unevenly penalize population subgroups (in particular, young, immigrants, high-income and residents living on the urban fringe).

Then, regarding the implementation of free-fare PT schemes, such systems can be also coupled of a higher access to other cultural activities (through a pass, as it is the case in Tallinn), adding social integration to the equity properties of the scheme. Indeed, transport equity can be favorably influenced by a *correct* distribution of accessibility over households (unequal accessibility is inevitable since space *by its very nature* is divided into center and periphery; see [47]), which can be obtained through the FFPT scheme. Changes in accessibility levels are often accompanied by changes in travel patterns and, in the longer run, by changes in land use (e.g.

dwelling price increase following from more affordable public transit services), with substantial feed-back impacts on accessibility levels (expulsing the most vulnerable to the outskirt of the city; see [35]) that need to be controlled. At least, one can also add that FFPT introduces a non-discriminatory benefit to all public transport users, regardless of their income level [11].

Acceptability Challenges

Experiencing free (or largely reduced) public transit fares can be framed as a trial period in order to secure public acceptability and break deadlock situations [38]. Thøgersen [61] shows for the case of Copenhagen that a public transport's monthly subscription card given for free to car drivers led to successful results, with a higher modal share of public transport even after the withdrawal of the scheme.

2.4 Conclusion

In this chapter, we shown that implementing pure regulatory tools or modifying the generalized cost of car trips through congestion tolling, parking charging or public transit faring all influence modal split on a different manner—by the fixed or variable (time, area-dependent, etc.) feature of the scheme considered, the volume of trip-makers involved, and over a different time-period.

In addition, the presented tools have dissimilar distributive effects on the social welfare, some being known for being more regressive than other.

At least, when acceptability issues are considered, psychological factors dominate self-interest variables. This makes the framing of the tool and the ordaining of the branded arguments the major factor for accepting a policy measure and thus for its successfully implementation.

References

1. Agence De l'Environnement et de la Maîtrise de l'Energie—ADEME (2011) Etat de l'art sur le développement des LEZ en Europe. Service Evaluation de la Qualité de l'Air, mise à jour mars 2011
2. Ajzen I (1991) The theory of planned behavior. Organ Behav Hum Decis Process 50:179–211
3. Armelius H, Hultkrantz L (2006) The politico-economic link between public transport and road pricing:an ex-ante study of the Stockholm road-pricing trial. Transp Policy 13 (2006):162–172
4. Arsenio E, Di Ciommo F, Dupont-Kieffer A, Fearnley N, Julsrud E, Kaplan S, Keseru I, Madre J-L, Martens K, Mitsakis E, Monzon A, Paez A, Papanikolaou N, Sauri S, Shiftan Y, Sivakumar A, Vallee D (2014). Transport equity analysis: assessment and integration of equity criteria in transportation planning. COST Action TU 1209 2014

5. Bamberg S, Fujii S, Friman M, Gärling T (2011) Behaviour theory and soft transport policy measures. Transp Policy 18(1):228–235, Jan 2011
6. Beck MJ, Rose JM, Hensher DA (2013) Consistently inconsistent: the role of certainty, acceptability and scale in choice. Transp Res Part E 56(2013):81–93
7. Bioethanol for Sustainable Transport—BEST (2010) The BEST experiences with bioethanol buses. BEST WP2 Deliverable No 2.08, Buses Final Report March 2010, Stockholm
8. Bonsall P, Young W (2010) Is there a case for replacing parking charges by road user charges? Transp Policy 17(2010):323–334
9. Bresson G, Joyce D, Madre J-L, Pirotte A (2003) The main determinants of the demand for public transport: a comparative analysis of England and France using shrinkage estimators. Transp Res Part A 37(2003):605–627
10. Button K (2006) The political economy of parking charges in "first" and "second-best" worlds. Transp Policy 13(2006):470–478
11. Cats O, Reimal1 T, Susilo Y (2014) Public transport pricing policy—empirical evidence from a fare-free scheme in Tallinn, Estonia. Paper submitted for presentation at the 93rd Annual Meeting of the Transportation Research 19 Board, Washington, Jan 2014
12. Commissariat Général au Développement Durable—CGDD (2011) Consommation de carburant: effets des prix à court et à long termes par type de population. Etudes et Documents n°40 avril 2011
13. Commissariat Général au Développement Durable—CGDD (2012) Etude sur les externalités des transports—Le monde routier. Service de l'Economie, de l'Evaluation et de l'Intégration du Développement Durable, Sous-Direction Mobilité et Aménagement
14. Commissariat Général à la Stratégie et à la Prospective—CGSP (2013) L'évaluation socioéconomique des investissements publics. Rapport de la mission présidée par Emile Quinet, septembre 2013
15. Charles L, Roussel I, Gobert J, Blanchet A (2011) Les initiatives ZAPA: un tournant dans l'action de la prévention de la pollution atmosphérique? Pollution atmosphérique n°210, avril-juin 2011
16. Charleux L (2014) Contingencies of environmental justice: the case of individual mobility and grenoble's low-emission zone. Urban Geogr 35(2):197–218
17. Crozet Y, Lopez-Ruiz HG (2013) Macromotives and microbehaviors: climate change constraints and passenger mobility scenarios for France, Transport Policy 2013 29 (C):294–302
18. De Borger B, Proost S (2012) A political economy model of road pricing. J Urban Econ 71 (2012):79–92
19. Dietz S, Atkinson G (2005) Public perceptions of equity in environmental policy: traffic emissions policy in an English urban area. Local Environ 10(4):445–459, Aug 2005
20. Ding C, Song S, Zhang Y (2008) Paradoxes of traffic flow and economics of congestion pricing. UNR joint economics working paper series working paper no. 08-007
21. Eliasson J (2008) Lessons from the Stockholm congestion charging trial. Transp Policy 15 (2008):395–404
22. Ellison RB, Greaves SP, Hensher DA (2013) Five years of London's low emission zone: effects on vehicle fleet composition and air quality. Transp Res Part D 23(2013):25–33
23. Eriksson L, Garvill J, Nordlund A (2006) Acceptability of travel demand management measures: the importance of problem awareness, personal norm, freedom, and fairness. J Environ Psychol 26(2006):15–26
24. European Commission (2011) White paper: roadmap to a single European transport area— towards a competitive and resource efficient transport system. COM (2011) 144 final, Brussels, 28 Mar 2011
25. European Commission (2012) EU transport in figures. Statistical Pocketbook 2012
26. European Environment Agency—EEA (2011) Transport infrastructure investments (TERM 019)—Assessment published Jan 2011
27. European Environment Agency—EEA (2012) EU transport in figures. Statistical Pocketbook 2012, © European Union 2012

28. Farber B, Li P, Nurul H (2014) Social equity in distance based transit fares. Paper presented to the 93rd annual meeting of the transportation research board, Washington, 12–16 Jan 2014
29. Festinger L (1957) A theory of cognitive dissonance. Standford University Press, Standford
30. Finn B (2005) Study of systems of private participation in public transport. PPIAF, Stockholm
31. Fu M, Andrew Kelly J (2012) Carbon related taxation policies for road transport: efficacy of ownership and usage taxes, and the role of public transport and motorist cost perception on policy outcomes. Transp Policy 22:57–69, July 2012
32. Gálvez T, Jara-Díaz SR (1998) On the social valuation of travel time savings. Int J Transp Econ 25(2):205–219
33. Givoni M, Banister D (2013) Moving towards low carbon mobility. Edward Elgar Publishing, Cheltenham, 1 Jan 2013, p 293
34. Gunn LA (1978) Why is implementation so difficult? Manage Serv Gov 33:169–176
35. Hansen WG (1959) How accessibility shapes land use. J Am Inst Planners 25(2):73–76
36. Hensher DA (2007) Bus transport: economics, policy and planning. research in transportation economics, vol 18. Elsevier, Amsterdam, p xix–xxviii, 1–507
37. Hivert L, Wingert J-L (2010) Automobile et automobilité : quelles évolutions de comportements face aux variations du prix des carburants de 2000 à 2008?. 68 pages, Chapitre de l'ouvrage collectif «Pétrole Mobilité CO2» coordonné par Y. Crozet, LET, pour PREDIT-DRI, juin 2010
38. Institute of Transport Economics—TØI (2011) How to manage barriers to formation and implementation of policy packages in transport. Deliverable 5, June 2011
39. International Transport Forum—ITF (2008) Transport infrastructure investment options for efficiency. OECD International Forum, OECD publishing, 14 Feb 2008
40. Ison S, Rye T (2003) Lessons from travel planning and road user charging for policy-making: through imperfection to implementation. Transp Policy 10(2003):223–233
41. Jaensirisak S, Wardman M, May AD (2005) Explaining variations in public acceptability. J Transp Econ Policy 39(2):127–153, May 2005
42. Kaufmann V, Guidez J-M (1996) Les citadins face à l'automobile—les déterminants du choix modal. Paris, Fonds d'Intervention pour les Etudes et Recherches, p 188 (rapport du FIER n °19)
43. Kilani M, Proost S, van der Loo S (2013) Road pricing and public transport pricing reform in Paris: complements or substitutes? Article in press
44. Kottenhoff K, Brundell Freij K (2009) The role of public transport for feasibility and acceptability of congestion charging—the case of Stockholm. Transp Res Part A 43 (2009):297–305
45. Li Z, Hensher DA (2012) Congestion charging and car use: a review of stated preference and opinion studies and market monitoring evidence. Transp Policy 20(2012):47–61
46. Madre J-L, André M, Rizet C, Leonardi J, Ottmann P (2010) Importance of the loading factor in transport CO2 emissions. 12th WCTR, Lisbon, Portugal, 11–15 July
47. Martens K (2011) Substance precedes methodology: on cost–benefit analysis and equity. Transportation 38:959–974
48. Martens K (2012) Justice in transport as justice in accessibility: applying Walzer's 'Spheres of Justice'. Transportation. doi 10.1007/s11116-012-9388-7, Springerlink.com
49. May AD, Kelly C, Shepherd S (2006) The principles of integration in urban transport strategies. Transp Policy 13(4):319–327
50. Meurisse B, Papaix C (2013) Overview of the policy toolbox for low-carbon road mobility in the European Union. Les Cahiers de la Chaire Economie du Climat, Série Informations et débats n°26
51. Nakamura K, Hyashi Y (2013) Strategies and instruments for low-carbon urban transport: an international review on trends and effects. Transp Policy 29(2013):264–274
52. Ortuzar JD, Willumsen LG (2011) Modelling transport, 4th edn. Mar 2011, ©2011 p 606
53. Parry I, Small K (2009) Should urban transit subsidies be reduced? Am Econ Rev 99 (3):700–724

54. Proost S, Van Dender K (2001) The welfare impacts of alternative policies to address atmospheric pollution in urban road transport. Reg Sci Urban Econ 31(2001):383–411

55. Raux C, Souche S (2001) L'acceptabilité des changements tarifaires dans le secteur des transports : comment concilier efficacité et équité?. Revue d'économie régionale et urbaine, n °4, pp 539–558

56. Santos G, Behrend H, Maconi L, Shirvani T, Teytelboym A (2010) Part I: externalities and economic policies in road transport. Res Transp Econ 28(2010):2–45

57. Schade J, Schlag B (2003) Acceptability of urban transport pricing strategies. Transp Res Part F 6(2003):45–61

58. Schipper L, Cordeiro M, Wei-Shiuen NG (2007) Measuring the carbon dioxide impacts of urban transport projects in developing countries. World Resources Institute, 15 Nov 2007

59. Schuitema G, Steg L (2008) The role of revenue use in the acceptability of transport pricing policies. Transp Res Part F 11(2008):221–231

60. Shoup D (1999) The trouble with minimum parking requirements. Transp Res Part A: Policy and Pract 33(7–8):549–574, Sep–Nov 1999

61. Thøgersen J (2009) Promoting public transport as a subscription service: effects of a free month travel card. Transp Policy 16(6):335–343

62. Verhoef ET (2005) Second-best congestion pricing schemes in the monocentric city. J Urban Econ 58(2005):367–388

63. Verhoest K, Carbonara N, Lember L, Petersen OH, Scherrer W, Van den Hurk M (2013) Public private partnerships in transport: trends and theory P3T32013 Discussion Papers Part I Country Profiles

64. Victoria Transport Policy Institute—VTPI (2013) Understanding transport demands and elasticities. How prices and other factors affect travel behavior? Todd Alexander Litman 2005–2013

65. World Bank (2009). Urban transport and CO_2 emissions: some evidence from Chinese cities. Working Paper—June 2009, Georges Darido, Mariana Torres-Montoya and Shomik Mehndiratta

66. Xenias D, Whitmarsh L (2013) Dimensions and determinants of expert and public attitudes to sustainable transport policies and technologies. Transp Res Part A: Policy Pract 48:75–85

67. Zatti A (2004) La Tariffazione Dei Parcheggi come Strumento di Gestione della Mobilità Urbana: Alcuni Aspetti Critici. Quaderni del Dipartimento di Economia e Territoriale dell'Università di Pavia, 5

68. 6-t (2013) Etude de mobilité d'un échantillon d'habitants afin d'amener de nouveaux usagers vers le train: Expérience de marketing individualisé. 6-t Bureau de recherche. Commanditaire : Conseil Régional de Picardie, Partenaire : Emploi80 (ETTI) dans le cadre de la clause sociale

Chapter 3
Recent Innovation in Last Mile Deliveries

Eleonora Morganti and Laetitia Dablanc

Abstract In cities and metropolitan areas, last mile deliveries are a key factor contributing to local economic vitality, urban life quality and attractiveness of urban communities. However, the freight transport sector is responsible for negative impacts, mainly with regards to congestion, CO_2 emissions and air and noise pollution. In order to improve efficiency and reduce adverse impacts, private companies as well as city planners and policy makers have designed initiatives to promote urban logistics innovations—organizational and technological—and introduce new policies. In this chapter, we provide an overview of innovative measures for last mile deliveries. We discuss innovation for delivery vehicles: new concepts, sizes and technologies (including electric and hybrid powered engines). We then detail recent innovations in parcel delivery services for e-commerce, focusing on pickup points and automated lockers. Finally, we present recent initiatives on urban food logistics and the deployment of Food Hubs as urban consolidation centers for perishable products.

Keywords Mile deliveries · Urban logistics · Food hubs

3.1 Introduction

The world's population is becoming increasingly concentrated in cities, and the urban population is expected to rise from 3.6 billion in 2011 to 6.3 billion by 2050 [34]. Currently in Europe around 77 % of population lives in urban areas [34] and this is expected to increase to 84 % by 2050. Urban economies are evolving rapidly and they are more dependent on transportation systems, with more frequent and customized deliveries [7]. This is due to the major changes that have taken place in cities. As an example, the size of store inventories has shrunk, and businesses are

E. Morganti (✉) · L. Dablanc
IFSTTAR, SPLOTT, Cite Descartes, 14-20 bd Newton, 77447 Marne la Vallée, France
e-mail: eleonora.morganti@ifsttar.fr

© The Author(s) 2014   27
A. Hyard, *Non-technological Innovations for Sustainable Transport*,
SpringerBriefs in Applied Sciences and Technology, DOI 10.1007/978-3-319-09791-6_3

increasingly supplied on a just-in-time basis. The number of different products sold has increased considerably, and collections change several times a year. Express and urgent deliveries have increased. E-commerce deliveries have developed considerably, etc.

Indeed, last mile deliveries are a key factor which contributes to local economic vitality, urban life quality, accessibility and attractiveness of local community [26]. However, freight transport generates negative externalities including traffic congestion, environmental issues, traffic accidents, and energy consumption related to urban traffic. In large cities, one fourth of carbon dioxide, one third of nitrate oxides, and half of the particulates that come from transport are generated by trucks and vans [21].

Due to the rapid urbanization and to the increasing movements of goods and people, cities in the world are facing challenging questions related to last mile logistics. In order to improve efficiency and reduce the adverse impacts of urban freight distribution, various stakeholders in cities and metropolitan areas have launched initiatives to enhance freight logistics systems, finance organizational and technological innovation and introduce new traffic regulations.

Cities' efforts to manage last mile problems include consultation with private companies, traffic and parking regulations, and, increasingly, the promotion of innovative city logistics schemes such as consolidation schemes, modal shift and clean vehicles. Private logistics companies and shippers are also committed to more efficient urban logistics. As an obvious example, small scale consolidating facilities such as pickup points and locker services have been implemented in metropolitan areas worldwide in order to reduce parcel home deliveries.

Important innovations in last mile deliveries are specifically important for (but not exclusive to) the parcel delivery sector, in order to respond to the increased fragmentation of shipments in the final segment of the supply chain resulting from the spread of e-commerce. On-line shopping has experienced a steady growth over the past decade, and, currently about 48 % on the European population shops online [15]. In practice, e-commerce increases the challenges related to product distribution to the end-consumers, with direct impact on logistics systems in urban and suburban areas, where traffic congestion and accessibility are crucial factors. Initiatives which promote parcel consolidation are thus preferred by practitioners and policy makers that aim at optimizing urban freight.

Innovative solutions have been identified for specific logistics chains. Food logistics for instance has been targeted in order to make delivery trips to urban food businesses from the wholesale produce market, identified as Food Hub, more efficient.

For the purpose of this chapter, the definition of last mile deliveries includes all goods movements generated by the economic needs of local businesses, i.e. all deliveries and pick-ups of supplies, materials, parts, consumables, mail, and refuse that businesses require to operate. It also includes home deliveries by means of commercial transactions. We consider neither private transport undertaken by individuals to acquire goods for themselves (shopping trips), nor through traffic (trucks passing through a city en route to another destination without serving any business or household in the city). These two kinds of transport generate a large amount of vehicle-kilometers [21] and are legitimate policy targets, but for the purpose of this chapter we are looking at the accommodation and improved management of freight transport and logistics activities directly serving the local economy.

In this chapter, we provide an overview of the main innovations in urban freight in recent years, mostly in Europe (Sect. 3.1). In the Sect. 3.2, we discuss urban freight research, as well as the diversity of logistics chains at the urban scale. Section 3.3 focuses on the environmental impact generated by urban transport schemes. In Sect. 3.4 we identify key innovations on last mile deliveries, i.e. pickup points as alternatives to home deliveries for parcels generated by e-commerce; Food Hubs and urban deliveries for food outlets; multimodal transport operations reducing the use of trucks in urban areas. Finally, concluding remarks are presented.

3.2 Urbanization and Effects on Freight Transport

Urban communities are growing rapidly and so do their economies, which offer constantly changing innovative services to attract people and businesses. The quality of life of city dwellers is strongly connected to the efficiency of the urban transport system, which has direct implications on the health of individuals, on local economy performance, and on local land uses.

Freight distribution is thus a key element to guarantee goods availability in metropolitan areas. The metropolitan areas, especially the largest ones, now serve as trade hubs and strongly rely on efficient freight movements. On the other hand, commercial transport contributes to environmental and social externalities generated by transport. In particular, the disquieting trends related to rising air pollution and congestion force local planners to devise strategies to reduce the negative impacts of transportation.

Policy response on city logistics issues is much needed, but the complexity of the urban system requires integrated multi-approach solutions. In modern metropolitan areas, freight planning and policies not only look at air quality but also deal with preserving economic performance for businesses and retailers, fostering individual and communal health, rationalizing the use of transportation, enhancing environmental friendly modes [26]. According to Dablanc and Rodrigue [10], the growing demand for urban freight activity, related to the functions of production, distribution and consumption, starts to be a serious issue in cities of more than one million inhabitants (in the US). Cities above four million should have specific management schemes on freight.

Worldwide, a growing number of cities are currently impacted by the negative aspects of urban goods movements. However, to identify successful solutions, viable from an economic and environmental point of view, is not an easy task for city planners. Two main reasons contribute to explaining the delay in implementing consistent city logistics measures: (i) urban freight transport is a very complex and heterogeneous system, involving a large variety of actors and sectorial supply chains and evolving fast in response to customers and business needs, and (ii) little is known about freight distribution and there are limited available data describing the variety of logistics chains taking place in modern cities.

3.2.1 The Variety of Urban Supply Chains

In a single city, vehicles, delivery times, and the size of shipments may differ for each business or customer. Freight transport—unlike passenger transport, which can be broken down into a few categories according to trip mode and purposes—is notable for its phenomenal diversity. In this wide array of supply chains, it is relevant to identify the main urban logistics chains and provide basic logistics and organizational characteristics that help depict the complexity of the freight distribution system (types of vehicle, delivery frequency, etc.). As shown in Table 3.1, the most important freight chains dedicated to consumer-related distribution are represented by independent retailing, chain retailing, food markets, parcels and home deliveries. Two additional logistics chains, serving producer-related distribution, impact the urban system, i.e. materials for building sites and waste collection and disposal activities.

3.2.2 Urban Freight Data

Urban goods movements account for 15–30 % of all vehicles-kilometers in cities [9]. However, in many cities, accurate statistics are not available and data on freight are much harder to find compared with information on passenger mobility. The scarcity of data on commercial vehicle traffic represents a key challenge for city planners committed to innovative urban freight systems [18]. Important steps have been taken over the last few years in different parts of the world, and in Europe urban freight has been the subject of major programs providing statistics and impact assessments. Within Europe, though, comparisons are difficult because of differing local survey methodologies. Surveys are also conducted differently overtime, making historical analyses and projections difficult. According to Dablanc [8] using a variety of sources, including LET [21] and Figliozzi [16], data on urban freight resulting from different studies converge. A city in a developed country generates about:

- one delivery or pick-up per employment job per week;
- 300–400 truck trips per 1,000 people per day;
- 30–50 tons of goods per person per year.

Three to five percent of urban land is devoted to freight transport and logistics. A city not only receives goods, but also ships them: 20–25 % of truck-kilometers in urban areas are outgoing freight, 40–50 % are incoming freight, and the rest both

Table 3.1 Main urban logistics chains

Logistics chain	Main characteristics
Independent retailing, including the informal sector and local convenience stores	Use of vans (or bikes and carriages in poorer countries) Large share of own-account Deliveries frequency: three to ten times a week, 30–40 % of all daily deliveries in a city
Chain retailing and commercial centers	Trucks and vans, better-loaded, larger share of consolidated shipments compared to independent retailing Transport providers and third party operators Delivery frequency varies according to the size of store
Food markets	Very diverse supply modes (including bicycles, and hand- or animal-driven carts in developing countries) Large share of own-account Few data exist on the actual volume and of freight flows and frequency
Parcel transport (less than a full truck load) and express services	Use of large vans or small to medium-sized trucks, based on consolidated delivery tours departing from cross-dock terminals in inner suburban areas Predominance of express transport companies (UPS, DHL, TNT, FedEx), very high number of contractors (small companies, owner-drivers)
Sub-sector: home deliveries	Large postal operators and new players specialized in parcel transport for e-commerce Increase in delivery options: scheduled delivery, lockers, pickup points, etc.
Building sites	Use of heavy trucks Multiple suppliers High number of deliveries, queuing, and general disorder on sites: poorly planned delivery schedules It represents up to 30 % of tons carried in cities
Waste collection and disposal	Use of ad hoc heavy trucks Specialized operators Daily activities

Source authors' elaboration on [9]

originates from and is delivered within the city [20]. A recent survey made in the Paris region [28] showed that there were 4.1 million deliveries and pick-ups every week in the region.

3.3 Last Mile Deliveries and Trucks and Vans

In large cities of developed countries, trucks and vans are the main transport mode for urban freight distribution operations. In particular, vans and light goods vehicles (LGVs) of 3.5 tons or under have been adopted for urban freight transportation because of the characteristics of final shipments: small deliveries, short trips with many stops, and a high spatial dispersion of the receivers [18]. Heavy trucks are still circulating in city centers, mostly early in the morning, to deliver goods to grocery retailers and commercial centers.

3.3.1 Environmental Impact of Urban Transport

Freight transport is responsible for a third of transport-related NOx and half of transport-related particulate matter [21]. Greenhouse gas emissions and noise pollution are also among the most severe environmental effects of urban freight transport. Along the supply chain, negative externalities generated by last mile operations are significant, due to the specific features of urban transportation fleets.

Urban freight vehicles can be quite old. In Dublin in 2004, a fourth of all vehicles were manufactured in or before 1994. Only 15 % of vehicles were new (one year or less). In the Milan region, 40 % of circulating trucks are more than ten years old (sources and other examples provided in [8]). The renewal of the freight fleet is generally slower than for non-urban road freight traffic, because urban freight involves numerous competing small operators that cut costs as much as possible [8]. Another important issue is road safety. Trucks are involved in a relatively small share of road accidents in cities, but these accidents are usually more serious (resulting in deaths). On London's roads in 2005, about 14 % of all collisions involving goods vehicles result in serious or fatal injuries, which is higher than the figure for other road users [4]. The conciliation of truck traffic with bicycle use has been a recent policy target in Paris and London following fatal collisions that received a lot of media attention.

3.3.2 Light Commercial Vehicles' Performance

In European cities, LGVs are of ever-greater importance for final deliveries and their activity has increased substantially overtime [5]. This has come as a result of the suitability and versatility of LGVs for a wide range of goods and servicing tasks. Access restrictions in city centers (such as the lorry ban in London) have put added pressure on trucking companies to convert to light vehicles. These vehicles provide many advantages in terms of logistics and organizational performance, but the reduced load capacity generates a higher number of trips, and thus, has a negative impact on urban congestion and air quality.

Air pollutant emissions vary according to engines. Diesel fueled vehicles generally are more fuel-efficient than comparable petrol engine vehicles, however they release higher quantities of pollutant emissions (carbon monoxide, oxides of nitrogen and particulates), (see Table 3.2 from [5]). Diesel LGVs are responsible for a disproportionate amount of pollution compared with their mere physical presence.

Table 3.2 LCVs pollutant emissions—petrol and diesel engine—g/km (Index: car without three-way catalyst pre-1993 = 100)

Type of VUL and Year	Carbon monoxide	Hydro-carbons	Oxides of nitrogen	Particulates	Carbon dioxide
Petrol LCV					
pre1994	136	96	94	19	111
1994–1997	20	3	19	2	140
1998–2001	5	2	16	1	143
2002–2005	4	1	7	1	136
2006	3	1	5	1	128
Diesel LCV					
pre-1994	10	19	81	187	143
1994–1997	5	9	63	51	143
1998–2001	5	9	60	53	143
2002–2005	3	7	45	37	131
2006	3	4	23	24	122

Note Petrol LGVs pre-1994 were without three way catalysts. Petrol LGVs have had three way catalysts since 1994
Source DfT 2008 in [5]

In 2012, 1.1 million new vans entered the EU market and approximately three quarters of these vehicles were sold in France, Germany, Italy, Spain and the United Kingdom. Of the newly registered vehicle fleet, diesel vehicles represented 97 % [14]. For the past decade, the EU fleet has experienced dieselization, with a 17 % increase of all diesel vehicles during 1996–2006. Just over 1 % of newly registered vehicles used liquid petroleum gas (LPG) or natural gas (CNG) and pure electric vehicles represent 0.5 % of the vehicles sold. These vehicles had on average 10–15 % lower emissions than diesel vans [14].

The supply of electric and hybrid LGVs is still very limited. Many manufacturers, however, are developing new models and now propose a larger range of vehicle solutions for urban freight distribution, including electrically assisted cargo cycles, plug-in electric quadricycles and ultra-light delivery vans. Currently, full electric vehicles represent 0.5 % of the vehicles sold. Positive trends in sales have been recorded in different European countries. In France, where public subsidies encourage the acquisition of electric and hybrid vehicles, they currently account for 1.4 % of light commercial vehicles [25].

Urban freight, featuring short trips and numerous deliveries, is particularly adapted to electric vehicles technology. Electric vehicles emit zero exhaust emissions and make little noise, thus their deployment in urban environment seems to generate direct positive impact on air quality improvement. However, their production process and indirect emissions from power generation has to be considered when promoting electric mobility.

Reduced emissions from road transport are an important factor in improving air quality in urban areas, particularly because the number of diesel vehicles has been increasing in many parts of the EU. Policy planners are thus committed to foster a shift to environmentally less-damaging modes of transport and technological improvements in the performance of road vehicles, such as gains in fuel efficiency and catalytic converters. At the end of 2013, the European Parliament [13] adopted new regulations setting carbon dioxide (CO_2) emission performance standards for light commercial vehicles (Regulation (EC) No. 510/2011). For vans, the mandatory target is 175 g of CO_2/km by 2017 and 147 g by 2020. This compares with an actual average of 203 g in 2007 and 181.4 g in 2010.

3.4 City Logistics Innovations

Many experiments in last mile deliveries have been designed and carried out in cities across the world. In particular, European cities developed many urban freight initiatives [31], and identified different pilot projects which can contribute to reducing the negative effects generated by the use of trucks and vans in urban areas, without decreasing the economic performance of local economies.

Cities in Europe are diverse, but share some commonalities that generate specific urban freight characteristics and challenges. In historic centers, a significant proportion of roads are narrow, with poor vehicle access and constrained loading/unloading facilities. Over the past decade, many city centers have undergone a "retailing renaissance," including groceries, and currently a large number of shops, businesses, and private receivers demand for last mile shipments. However, existing transport infrastructure is unable to cope with the resultant increase in freight. In many European cities, freight transport related air pollution has increased, one of the reasons being that, for financial reasons, many operators tend to use old trucks and vans to respond to the increasing demand for urban deliveries. Diesel trucks remain a major source of particulate matter (PM_{10}) and nitrogen dioxide (NOx). However, a complete phasing out of leaded gasoline, gradual abandonment of old diesel vehicles (see low emission zones below) and the promotion of cleaner fuels and electric vehicles open the way for a reduction in pollutant emissions from commercial vehicles in the years to come, notably in cities such as London, Gothenburg, Milan or Berlin. Traffic congestion remains a significant operational problem for the urban freight system.

Urban freight strategies with city logistics innovations can be implemented to respond to specific events or situations, as the London 2012 Olympics and Paralympics Games. The municipality was primarily concerned with coordinating and planning all freight-related activities, such as improving road safety, re-timing deliveries and collections, optimizing trip planning [33]. This event fostered the implementation of innovative services for city logistics such as delivering customers directly to their hotels. London has also designed consolidation schemes for building materials on major construction sites. The London Construction

Consolidation Centre was implemented in 2006 with funds from Transport for London and private investors. An assessment showed that the scheme achieved a 68 % reduction in the number of vehicles and a 75 % reduction in CO_2 emissions [32]. More recently, in London, large retailers promoted experiments on parcel delivery for e-commerce. Vans deliver groceries for commuters who order on line, allowing them to pick up their goods after work in selected car parks at metro stations.

Projects focused on shared city logistics services have been implemented in various European cities. In the Netherlands, Binnenstadservice.nl (BSS) is a network of small consolidation centers available to transport operators and shop-keepers to handle transport in the last mile distribution in cities. It started in April 2008 in the city of Nijmegen. By bundling deliveries from multiple suppliers, BSS offers a service to the small store-owners located in the inner centers, which usually ask for small and frequent shipments and do not have an optimized delivery scheme. According to Rooijen and Quack [27], the number of trucks and also the amount of truck-kilometers in the city centre decreased after the first year of operation in Nijmegen. The following years, the reduction in truck-kilometers increased progressively with the increase of stores affiliated to the project. The effects on local air quality and noise are limited, due to the amount of remaining freight, passenger and bus traffic and the high natural background concentration for PM_{10} and NOx, however a significant impact has been reached in terms of traffic safety and reduced congestion. BSS-franchise is now replicated in ten Dutch cities, such as Maastricht, Utrecht and Rotterdam.

Trends such as e-commerce and retailing renaissance in city centers add further challenges for urban freight transport on overall mobility. Distribution sector operators look for fast, customized shipments services to keep and gain customers. Here we present innovations related to (i) parcel delivery and home delivery services; and (ii) goods deliveries for independent retailers in the food sector.

3.4.1 E-Commerce and Alternatives to Home Delivery

Over the past 10 years, the spread of online shopping has generated significant demand for dedicated delivery services to the end consumer. This has resulted in the increasing fragmentation of shipments in the last mile, affecting the management of the final segment of the supply chain [11, 29]. Consequently, e-commerce increases the challenges facing product distribution, with direct effects on logistics systems in urban and suburban areas where traffic congestion and accessibility are crucial factors. Home deliveries constitute the most problematic solution for e-commerce deliveries in terms of service costs and organization [30].

Although home deliveries are usually preferred by online shoppers [6], we are seeing the development of alternative solutions which satisfy both consumer demand for flexibility and firms' need to optimize parcel distribution through consolidated shipments. In Europe, automated parcel stations (APS) equipped with

Table 3.3 Trends for reception point networks in Europe

Company	Service type[a]	Country	No. sites 2008	No. sites 2012	Growth rate 08–12	Parcel volumes 2012
ByBox	APS	UK	1,000	1,300	+30 %	N.A.
Collect Plus	PP	UK	Not available	5,000	N.A.	N.A.
PackStation	APS	Ger.	1,000	2,500	+150 %	N.A.
Paketshop (Hermes)	PP	Ger.	13,000	14,000	+7.7 %	N.A.
ByBox	APS	F	Not implemented	170	N.A.	N.A.
Cityssimo	APS	F	20	33	+55 %	N.A.
Kiala	PP	F	3,800 (with M.R.)	4,500	+18 %	15 million
Pickup Services	PP	F	3,100	5,200	+68 %	9 million
Mondial Relay (Point Relais)	PP	F	3,800 (with Kiala)	4,300	+13 %	12 million
Relais Colis (Sogep)	PP	F	4,000	4,200	+5 %	23 million

[a] *APS* automated pack station; *PP* pickup point
Source [23]

lockers and pick-up points (PP), which are stores providing parcel drop-off and pick up, are fast-growing solutions. These two end-delivery options are playing a decisive role in the reorganization of commercial and logistics activities [1] and are becoming key features of the strategy of e-commerce and transport players.

Alternative delivery networks have recently developed in all European countries, especially in northern Europe where e-commerce and delivery services are more mature than in the rest of Europe. As reported in Table 3.3, strong trends towards an intensification of reception point networks are observed in the United Kingdom, in Germany and in France, namely Europe's largest online markets, which together represent 71 % of European e-commerce with revenues amounting to €143.2 billion in 2011 [18]. Currently the largest APS network is the Packstation network operated by DHL/Deutsche Post in Germany (2500 locations around the country). Locker box networks have a limited presence in France, as witnessed by the very small network of 33 kiosks run by La Poste under the name of Cityssimo. New operators such as ByBox (originally from the UK) and Neopost are likely to extend these services in Europe in the coming years.

In some countries, PP networks experienced a large increase. Postal operators traditionally provide this option delivering parcels in postal offices. For example, the Swedish operator PostNord provides about 5,000 distribution points to the end consumer in Sweden, Norway, Finland and Denmark. This delivery service is no longer exclusively provided by postal operators, in fact a growing number of transport and shipping operators have created their own network of affiliated shops and stores, where parcels are delivered waiting for e-shoppers to pick up the goods ordered on-line.

In the US, the online giants Amazon and Google (Google has opened an internet sales platform similar to Amazon's market place) recently decided to invest in their own branded locker box solutions and are in the process of deploying pilot pickup/ drop off sites. Similarly, in France new players are emerging and new partnerships set up, such as the takeover of the Kiala PP network by UPS in February 2012, and the takeover of the Pickup Services PP network by the French company La Poste, via its subsidiary GeoPost in 2009. The continuing influx of newcomers to the end-delivery sector shows that the market has not reached saturation, and there is substantial demand for new options for flexible and quick deliveries. Large retailers, for example, create ad hoc reception points located in their own stores, or in specific facilities where there are no points of sales in the vicinity. French grocery retailers increasingly provide this service, called *"Drive"* (even in French), enhancing on-line sales for food products.

3.4.1.1 Pickups Point Networks in France

PP delivery systems represent a traditional solution developed in Europe when distance selling was made by way of printed catalogues. In the era of e-commerce and multi canal distribution, they have been automated through sophisticated ICT tools in order to track and trace parcel shipments and provide a professionalized and streamlined process better adapted to e-consumers' demands. The French scenario is especially interesting because of the wide range of operators and their comprehensive deployment across the country. In France, four competing providers are growing rapidly and managing increasingly large volumes of parcels. Today 60 million parcels (about 20 % of shipments resulting from online shopping) are delivered through a PP instead of at home [23]. These PP providers—Mondial Relay, Relais Colis, Kiala and Pickup Services—have developed standardized delivery solutions for the whole country and in 2013 each of the networks provided access to a pickup point in under 10 min by car or on foot (depending on the area) to 90 % of the French population.

PP networks have experienced strong growth: the aggregate number of ventures serving as PP rose from 10,900 in 2008 to 18,200 in 2012, i.e. an increase of 67 %. The French system of *point relais* (reception points) has atypical features, such as its early development and the large number of players, with different shareholding structures. As presented in the introduction, there are four competing PP network operators in France (Mondial Relay, Kiala, Relais Colis and Pickup Services), and the development of their networks is fairly similar (see below). These providers are medium-sized, whereas in most countries the market is dominated by one or two large operators (e.g. Hermes in Germany, which is almost the country's only PP network operator, in parallel with DHL Packstations providing APS).

Table 3.4 Pick-up point density over population and e-shoppers, France 2008–2012

Firm	PP per 100,000 inhabitants 2008	PP per 100,000 inhabitants 2012	PP per 10,000 e-shoppers 2008	PP per 10,000 e-shoppers 2012
Kiala	6.1	7	1.7[a]	1.4
Pickup Services	5	8.1	1.4	1.7
Point Relais (Mondial Relais)	6.1	6.3	1.7[a]	1.3
Relais Colis	6.4	6.6	1.8	1.4
Average	5.9	7	1.6	1.5

[a] Mondial Relay and Kiala shared the network until 2012

The initial rise of PP operators in France derives from the development of mail-order selling during the 1980s [2]. Sogep—known as Relais Colis—and Mondial Relay were created by two mail-order companies, respectively La Redoute and 3Suisses, with the aim of improving the efficiency of their shipping services. These operators expanded their networks during the 1990s, driven by a sequence of postal strikes, and are now among the biggest players on the French market. The spread of e-commerce opened the way for two additional PP companies, the Belgian firm Kiala and Pickup Services, a French start-up created in 2004. The rise of these companies has not gone unnoticed by the major delivery and transport players, such as UPS and La Poste, which have shown particular interest in the IT system and e-logistics data networks set up by the two firms. As mentioned earlier, UPS and La Poste have bought Kiala and Pickup Services respectively. These networks have a quite similar spatial deployment and standard of service across France. Each of the operators provides online shoppers with between 4,000 and 5,500 pickup points across the country, i.e. a network which is almost a quarter of the size of the network of post offices. In total, the number of pick up points in France is as high as the number of post offices.

Table 3.4 presents changes in the density of PPs for each network, showing the increase in points available to the French population between 2008 and 2012. The effect of the growth in e-commerce activity is that each PP site is now serving a larger number of online shoppers. The risk is thus related to the potential high increase in parcel volumes. Networks could reach saturation and reduce the quality of their services.

The four current networks primarily rely on small independent local shops, such as florists, bars, tobacco shops and press kiosks. Shop-owners that decide to enter the network receive a small fee, ranging from €0.15 to 1.50 per parcel. The tasks include reception, storage, returns management and, in some cases, also payment of the items. Small retailers and shops agree to participate and offer pickup delivery services in order to attract potential customers and receive additional revenue. Side-effects of this logistics activity provided by commercial outlets can be customers

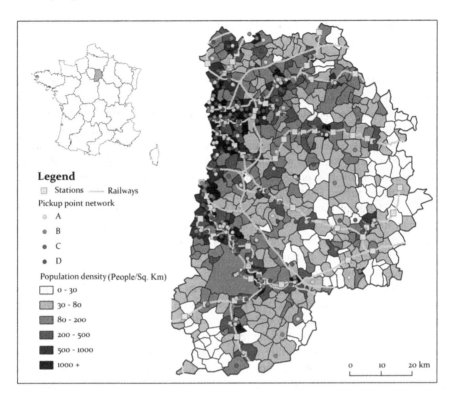

Fig. 3.1 Type of partner store in the pickup points network, Seine-et-Marne. Map by F. Fortin, IFSTTAR 2014

and e-consumers conflicts, and reduced selling area because of parcels storage. Figure 3.1 represents one example of PP deployment in France, looking at the example of the Seine et Marne Department, immediately East of the Paris dense metropolitan area. It shows that PP networks present similar locational patterns and tend to target the same areas, i.e. the most populated areas, where there is a very large number of potential partnering shops. Moreover, a large number of PP sites are located near commuter railway stations: one train station out of two in Seine-et-Marne is within a 300 m range from a PP [23]. Shops nearby railway stations are thus targeted as priority sites to recruit stores to be included in the network.

The deployment of PP and APS proves that these alternatives to home deliveries are successful solutions in order to reduce the risk of missed delivery and the highly fragmentation of parcel flows generated by e-commerce. In highly populated areas, where most people walk or use public transportation, these options are expected to generate high impact in reducing shopping and delivery trips. However, in suburban and rural areas, the potential gains in reduced vehicle-kilometers by vans and LGVs may be lost due to increased private car trips of web-consumers picking up their parcels at the pickup point.

3.4.2 Food Hubs and Food Deliveries to Independent Retailers

Urban food strategies become targets of growing interest by city authorities, and their attention is now focused on food systems and their interconnections with other community systems, including transport, land use, and waste management. In a growing number of cities such as San Francisco, New York and London, policy makers are setting a variety of measures to better integrate food issues in the urban policy agenda, and, together with the community and supply chain actors, they are implementing innovative projects dealing with provisioning and distributing food. Most of these urban food strategies include significant connections with city logistics and freight transport issues, in order: (i) to reduce air pollution; (ii) to contribute to the enhancement of food access and quality in urban environment; (iii) to improve the resources efficiency and cost effectiveness of the transportation of goods, taking into account external costs.

Despite the growing interest in city logistics projects that include food products, these initiatives are still limited due to three main obstacles. First, food logistics imposes constraints that do not apply to non-food supply chains, especially for products that require cold chain technology. Other constraints, such as short lead times and specific handling procedures, as well as regulatory issues (related to temperature requirements), further increase the operational costs of delivery services and, therefore, limit the economic viability of coordinated food logistics projects. Finally the high number of suppliers and receivers directly affects the feasibility of coordinating urban transport operations.

Last mile logistics for fresh food products in middle-sized European cities presents inefficiencies [22]. The "last food mile" for independent retailers and food services located in city centers is characterized by small-scale goods distribution, from terminals and facilities usually located in a range of 50 km. Receivers require daily express deliveries in a limited number of parcels, within a narrow time-frame. Moreover, food deliveries are usually informal logistics activities, many by own account transport carried out by food suppliers, producers or even shop-owners themselves. Third-party logistics and transport operators are not always appropriate for these supply chains. With regards to the efficiency of transport operations, the dispersion of receivers within the urban area entails the use of a large number of commercial vehicles operating below their maximum carrying capacity, with a high incidence on empty runs. Most urban food deliveries are made by ways of old diesel vehicles, e.g. small trucks and vans of more than 10–15 years, which consume large quantities of fossil fuel, and also generate higher quantities of pollutant emissions such as nitrogen dioxide and particulate matter.

3.4.2.1 Wholesale Produce Markets as Food Hubs

Some innovations in "last food miles" identify the wholesale produce market as a Food Hub, and develop the consolidation functions which have historically always been a part of wholesale markets. In European cities, where available land for logistics activities is limited or nonexistent, the wholesale market place is usually located in a strategic area, close to the consumption basin and well-connected to the main road networks. The wholesale produce market plays the role of a network shaping agent in the urban food supply system, grouping local and regional (and then global) producers, and providing direct links with urban retailers.

Wholesale produce markets (WPMs) already play a relevant role in managing perishable goods flows. It is possible to enlarge the role of wholesale markets to full Food Hubs, intended as supply chain intermediaries playing a new role in the urban food provisioning system. Food hub organizational models focus on including environmental and social criteria, associated with sustainable food systems, into market management processes. Three core components characterize Food Hubs [3], such as:

1. Aggregation, distribution and wholesale;
2. Active coordination;
3. Permanent facilities together with additional logistics, marketing and communication services.

As shown in Fig. 3.2, the new organization of WPM performs a double function. By giving priority to current demands for market efficiency in order to create a sustainable food system, it covers the functions of a Food Hub, while the services it provides for sustainable logistics at the urban level cover the functions of an urban consolidation centre. The traditional role of WPMs has been warehousing (straightforward storage) but wholesale markets can also be seen as consolidation and cross-docking facilities, due to the importance of just-in-time deliveries for fresh food distribution chains.

Example of Food Hubs can be found in the UK, in Italy [22] and in the United States [3], where a growing number of wholesale markets are fostering logistics services to supply chain operators and to the community. WPMs became urban

Fig. 3.2 Wholesale produce market in the urban food supply chain [24]

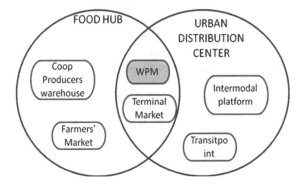

consolidation centres for the food supply chain, and they can be grouped with intermodal terminals (i.e. in Padua, Italy) and transit points (i.e. La Rochelle, France) for their logistics role. In practice, WPMs perform the functions of Food Hubs, by providing a market place for local producers and global food suppliers, and the functions of a consolidation centre, where transport operators consolidate parcels of produce to urban food businesses and organize the deliveries trips.

3.4.3 Cargo-Cycles and Waterways as Alternatives to Motor Vehicles

Several innovative projects on green urban freight have focused on alternative modes of transport. European cities are testing last mile delivery initiatives which combine bicycle and river transport. In Amsterdam, FoodLogica provides food final deliveries to final consumers and food outlets. In 2013, the network of food businesses (hotels, restaurants, cafés) potentially interested in the project was identified. Deliveries are made by a fleet of non-polluting transport modes (electric boats, trucks and bikes), identified as "Smart food grid:" a distribution network of zero-emission vehicles to bring local produce into the city [17]. In Paris, "Vert chez Vous" is a pilot project combining bicycle and river transport, which started in 2011. It is currently experiencing a steady growth in parcel volumes. The key element of this project is the double function of the boat, as river shuttle as well as as a moving logistics terminal. Electrically assisted cargo-bikes (tricycles with a large container in the back) are shipped by boat, loaded at 7 am at the city outskirts together with the parcels to be delivered to businesses located in the city center. While approaching the city center, the staff on board sort parcels by bike delivery route. The barge makes five stops along the Seine and, at each stop, one or two cargo-bikes are debarked and start their local delivery tour. Once the drop offs are completed, the bike joins the barge two stops further on and can be filled up again for another tour.

A third project using waterways for last mile deliveries has been implemented, again in Paris. The French food retail company Franprix, together with the logistics service provider Norbert Dentressangle and with support from local administrations, developed and implemented a multi-modal transport solution for supplying Franprix stores located in Paris. Previously, the stores were supplied by trucks directly from the company's distribution center 20 km away from Paris. Now, a container barge shipping 26 containers enters the city everyday at 5 am (the trip lasts 3 h) and drops off containers at the quai de la Bourdonnais, not far from the Eiffel tower. Trucks pick up containers (which are ready for deliveries) and provide the final delivery miles to about 80 supermarkets.

According to an ex ante assessment (cited by [11]), the use of the barge avoids the equivalent of about 450,000 vehicle-kilometres by trucks per year, when fully operational. This is equivalent to a 37 % reduction in CO_2 emissions for the trip between the regional warehouse and the stores.

3.5 Conclusion

In this chapter we presented last miles innovations implemented in different cities in Europe. These projects have a double objective: adapting city logistics schemes to the recent evolution in transport and service demand from urban communities; and reducing emissions and congestion generated by traditional delivery trucks and vans. Various logistics chains are targeted by urban freight initiatives, such as parcel deliveries from e-commerce or food deliveries from wholesale markets to urban outlets. In some cases, multimodal transport operations are promoted, mostly through local partnerships between public and private actors, in order to reduce the use of trucks in last mile deliveries and to favor more sustainable modes such as waterways and bi- or tri-cycles. Most of these projects' market areas are limited and the potential benefits—especially during startup phases—are limited compared with total urban freight flows. However, the implementation of city logistics initiatives represents a crucial element in the process towards a more sustainable urban transport system, paving the way for more substantial changes, with high potential benefits resulting from the adoption of new practices and new technologies by all stakeholders in urban supply chains.

References

1. Augereau V, Dablanc L (2008) An evaluation of recent pick-up point experiments in european cities: the rise of two competing models? In: Taniguchi E, Thompson R (eds) Proceedings of the 5th international conference on City Logistics, Nova Science Publisher, Inc., New York
2. Augereau V, Curien R, Dablanc L (2009) Les relais-livraison dans la logistique du e-commerce, l'émergence de deux modèles. Cahiers scientifiques du transport 55:63–96
3. Barham J, Tropp D, Enterline K, Farbman J, Fisk J, Kiraly S (2012) Regional food hub resource guide. U.S. Dept.of Agriculture, Agricultural Marketing Service. Washington, April 2012. http://dx.doi.org/10.9752/MS046.04-2012. Accessed 11 May 2013
4. Bestufs (2006) Quantification of urban freight transport effects I. Deliverable. www.bestufs.net. Accessed 10 Oct 2009
5. Browne M, Allen J, Nemoto T, Visser J (2010) Light goods vehicles in urban areas, procedia —social and behavioral sciences, 2, pp 5911–5919. http://hermes-ir.lib.hit-u.ac.jp/rs/bitstream/10086/22051/1/0101102301.pdf. Accessed 11 Nov 2013
6. CREDOC (2010) Le profil des acheteurs à distance et en ligne. Reed CCI, FEVAD, La Poste, p 62
7. Dablanc L (2007) Goods transport in large european cities: difficult to organize. Difficult to Modernize Transp Res Part A 41:280–285

 8. Dablanc L (2009) Freight transport, a key for the new urban economy, report for the world bank as part of the initiative freight transport for development: a policy toolkit. p 52
 9. Dablanc L (2013) City logistics. In: Rodrigue JP, Comtois C, Slack B (eds) The geography of transport systems, 2nd edn. Routledge, New York
10. Dablanc L, Rodrigue JP (forthcoming) The geography of urban freight. In: Giuliano G, Hanson S (eds) The geography of urban transportation, 4th edn. Guilford Press, New York
11. Depierre D (2013) Ha.Ro.Pa., une sinergie portuaire. Paper presented at the 2nd international Conference Chaire Unesco «Alimentations du monde», SupAgro, Montpellier France, 1st February 2013
12. Esser K (2006) B2C e-commerce impact on transport in urban areas. In: Taniguchi E, Thompson R (eds) Recent advances in city logistics. Elsevier, Amsterdam, pp 437–448
13. European Parliament (2011) Regulation setting emission performance standards for new light commercial vehicles as part of the Union's integrated approach to reduce CO2 emissions from light-duty vehicles (Regulation (EC) No. 510/ 2011). http://eur-lex.europa.eu/LexUriServ/LexUriServ.do?uri=OJ:L:2011:145:0001:0018:EN:PDF. Accessed 2 Feb 2014
14. European Environmental Agency (2014) Van manufacturers must make new models more efficient by 2020. http://www.eea.europa.eu/highlights/van-manufacturers-must-make-new. Accessed 12 Apr 2014
15. Eurostat (2013) Information society statistics http://epp.eurostat.ec.europa.eu/statistics_explained/index.php/Information_society_statistics_at_regional_level. Accessed 27 Feb 2014
16. Figliozzi MA (2007) Analysis of the efficiency of urban commercial vehicle tours: Data collection, methodology, and policy implication. Transp Res Part B 41:1014–1032
17. Food Logica (2013) www.foodlogica.com
18. Giuliano G, O'Brien T, Dablanc L, Holliday K (2013) NCFRP project 36(05) synthesis of Freight research in urban transportation planning, Washington: National Cooperative Freight Research Program. http://www.trb.org/Publications/Blurbs/168987.aspx. Accessed 6 Dec 2013
19. Kelkoo (2012) L'e-commerce en Europe. L'e-commerce transfrontière - Conférence Acsel, Paris, 26 Jan. http://press.kelkoo.co.uk/wpcontent/uploads/2012/01/25012012_Bilan-ecommerce-Acsel_-FINAL.pdf. Accessed 28 Jan 2013
20. LET (2000) Diagnostic du transport de marchandises dans une agglomération. DRAST/Ministère, Paris
21. LET (2006) Méthodologie pour un bilan environnemental physique du transport de marchandises en ville. ADEME/Ministère des Transports co-Publishers
22. Morganti E (2011) Urban food planning, city logistics and sustainability: the role of the wholesale produce market. The cases of Parma and Bologna food hubs PhD Dissertation, Alma Mater Studiorum, University of Bologna, p 211
23. Morganti E, Dablanc L, Fortin F, (2014). Final deliveries for online shopping: French operators'strategies according to the customers and the area they live in. Research in Transportation Business and Management, in press
24. Morganti E, Gonzalez-Feliu J (2014) Approvisionnement urbain des produits frais: le rôle du marché en gros dans la logistique urbaine. Presented at 51st conference of Association de Science Régionale de Langue Française (ASRDLF), Marne la Vallée, France, 7–9 July 2014
25. Observatoire du véhicule d'entreprise (2013) Immatriculation hybrides et électrique http://www.observatoire-vehicule-entreprise.com/immatriculations-hybrides-et-%C3%A9lectriques-en-octobre-2013. Accessed 12 Jan 2014
26. OECD (2003) Delivering the goods, 21st century challenges to urban goods transport, OECD Publishing, Paris. http://www.internationaltransportforum.org/pub/pdf/03DeliveringGoods.pdf. Accessed 22 Apr 2014
27. Rooijen T, Quack H (2010) Local impacts of a new urban consolidation centre—the case of Binnenstadservice.nl Innovations in City Logistics. Procedia Soc Behav Sci 2(3):5967–5979
28. Routhier JL (2014) Paris urban freight survey: first results. Presented at Metrofreight Special Session, Transport Research Arena, CNIT Paris La Defense, 17 April 2014

29. Schewel L, Schipper L (2012) Shop 'till we drop: a historical and political analysis of retail goods movement in the United States. Environ Sci Technol 46–18:9813–9821
30. Song L, Cherrett T, McLeod F, Wei G (2009) Addressing the last mile problem. Transport impacts of collection and delivery points. Transp Res Rec: J Transp Res Board 2097:9–18
31. SUGAR (2011) City Logistics Best Practices: A handbook for authorities. European Commission, INTERREG Program, SUGAR Project, final publication. p 272. www.sugarlogistics.eu/pliki/handbook.pdf. Accessed 20 Nov 2013
32. Transport for London (2009) London Construction Consolidation Centre Final Report. http://www.tfl.gov.uk/cdn/static/cms/documents/lccc-final-report-july-2009.pdf. Accessed 22 May 2014
33. Transport for London (2012) Delivering a road freight legacy. Working together for safer, greener and more efficient deliveries in London. http://www.tfl.gov.uk/cdn/static/cms/documents/delivering-a-road-freight-legacy.pdf. Last accessed 10 April 2014
34. United Nation (2013) Population, Development and Environment 2013 Economic and social Affair Publication

Chapter 4
New Innovative Rail Services: Stakes and Perspectives

Corinne Blanquart and Thomas Zéroual

Abstract *Background and aim*: According European transport policy, information literacy plays a crucial role to promote sustainable transport. The purpose of this article is to outline the European transport policy impact on logistics organization. To analyze this impact, we will focus on freight transport in this chapter. *Methods*: European regulatory bills analyses since 1991 will be used as the basis of the evaluation of the transport policies limits. *Results*: We conclude that current European transport policies have several limitations. First of all, non-technological innovation is less fostered. Secondly, transportation policies still over-emphasize industrial activities. Innovation analysis for service activities is not very common yet. Finally, shippers' needs are neglected. *Conclusion*: Technological impact fostering sustainability is over-estimated. By the way, transport policies should identify much more shippers' needs. These needs are differentiated. They involve much more than technological solutions in order to foster sustainable transportation.

Keywords Transport · Freight · Innovation · Sustainability · Evaluation

4.1 Introduction

Transportation's environmental impact is linked to both the supply of energy and the resulting emissions. Both the energy consumption and the emission of greenhouse gases have been increasing steadily in this sector.

C. Blanquart (✉)
IFSTTAR–SPLOTT, BP 317, 59666 Villeneuve d'Ascq Cedex, France
e-mail: corinne.blanquart@ifsttar.fr
URL: http://www.ifsttar.fr/linstitut/ame/laboratoires/splott/?police=yuihyfoacgrtdzsd

T. Zéroual
ESCE-CIRCEE, 10, rue Sextuis Michel, 75015 Paris, France
e-mail: thomas.zeroual@esce.fr
URL: http://www.esce.fr/recherches/recherche-a-esce/

© The Author(s) 2014   47
A. Hyard, *Non-technological Innovations for Sustainable Transport*,
SpringerBriefs in Applied Sciences and Technology, DOI 10.1007/978-3-319-09791-6_4

Compared to other transport modes, rail freight transport seems not to be the most environmentally damaging. Specifically, GHG emissions and energy consumption are much lower than the average of the transport sector.

In France, as in Europe, a sustainable transport policy is strongly needed because of the increase in environmental externalities (in terms of emissions and energy consumption).

The European Commission then redefined sustainable transport using the notion of co-modality, which can be defined as *"the efficient use of isolated or combined modes of transport"* [11]. Today the goal is to optimize the use of transport networks.

To support co-modality, command of information plays an important role for European public policy. In particular, improving the control of information is possible thanks to new management systems. Europe invests in "intelligent" freight, which means the transport of freight responds to increase needs of business information, internationalization, securitization and traceability of flows, as well as dematerialization of transportation procedure.

The results of political efforts have been mixed concerning transportation compared with other sectors [40]. Several factors can explain these results. In this chapter, we will explain why sustainable transport policy effects are limited.

4.2 Railroad Transportation and Sustainability

4.2.1 Transportation: A Sector with High Negative Externalities

Even though there is no direct causal relationship between transportation and economic growth, there is clearly one between transportation and negative externalities. These negative externalities are widely known and correspond to two types of impact. First of all, transportation's environmental impact is linked to both the supply of energy and the resulting emissions. Both the energy consumption and the emission of greenhouse gases have been increasing steadily in this sector (see Fig. 4.1).

If we focus on Europe-wide energy consumption in the transportation sector, we notice that no country has been able to reduce its consumption between 1995 and 2006 (see Fig. 4.2).

Second, transportation leads to a diverse negative social impact. For instance, working conditions have gotten worse (nightshifts, fewer and short breaks and more intensive physical work) as global competition increases [3].

These impacts are noted on a global level (greenhouse gases) and on a local one (urban traffic jams, noise, pollution, loss of time), as shown in Table 4.1.

Transportation thus imposes two types of externalities: environmental and social costs. The growth of transportation's negative externalities on the environment is more worrisome than that on social issues. A 2009 communiqué of the European Commission [7] stressed the need for effort on the environmental issue: *"it's principally in the area of environmental protection that further progress should be pursued"*. We shall therefore focus on this aspect in this chapter.

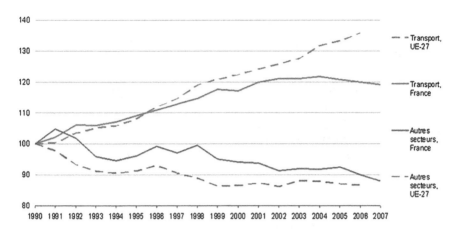

Fig. 4.1 Transportation and GGE. *Source* Citepa [4]

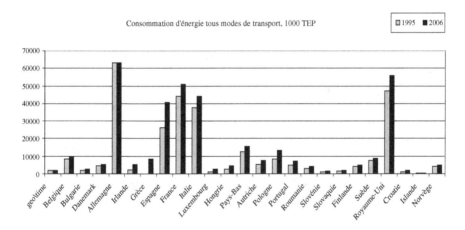

Fig. 4.2 Energy consumption from transportation needs in Europe. *Source* IFEN (2008)

4.2.2 Higher Negative Environmental Externalities in Freight than in Passenger Transport

Passenger transport causes more environmental damage than freight transport in absolute terms. However, the increase of freights environmental externalities is relatively higher than passenger, due to several factors.

First, the annual growth rate in freight transport is higher than in passenger transport. On a global level, the annual growth rate in passenger transport would have been multiplied by 3 from 1850 to 1990 for all modes of transport according to the Canadian Centre for Sustainable Transportation. Ceteris paribus, the annual growth rate in freight transport would have been multiplied by 1,000. The

Table 4.1 The negative external impact of transport at different scales

Social/ Environmental	Externalities	Local	Regional	National	UE	World
	Congestion	–	–	/	/	/
	Accidents	–	–	CNSR	–	–
	Space allocation conflict/ collision	–	–	–	–	/
	Atmospheric pollutions	–	–	–	/	/
	GHG emissions	/	/	PNLCC	Kyot	Kyoto
	Noise pollution	–	–	/	/	/
	Dividing effect	–	–	–	/	/
	Consumption of non-renewable resources	/	/	–	–	–
	Soil and subsoil pollution	–	–	–	/	/
	Waste generation	–	–	/	/	/

Source Savy [36]

conclusions are identical on the most recent period even if the proportions vary. Freight transport grew by 26 % in per-kilometres in Europe over the past decade where as passenger transport grew by only 2 % ceteris paribus.

Second, freight transport causes more environmental damage compared to others sectors. For instance, greenhouse gas emissions are higher in freight transport compared to the energy or industry sector.

Finally, commodity flows per-kilometres increased more than five times from 1945. This increase has mainly benefited the road sector, particular in regards to freight transport. The market share of the railway mode has decreased continuously since the 1950s.

4.2.3 Rail Freight Transport: Environmental Externalities and Growth Rate

Compared to other transport modes, rail freight transport seems not to be the most environmentally damaging. Specifically, GHG emissions and energy consumption are much lower than the average of the transport sector. Regarding the GHG emissions, the rail freight share decreased by 45 % between 1990 and 2006 (Figs. 4.3 and 4.4).

Many studies confirm this lower growth, as shown in projections in chart 4.

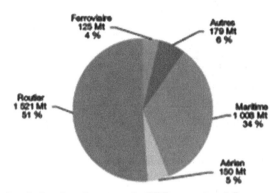

(en Mt de CO₂ et % des émissions du fret)

Source : CAS d'après les données 2005 du FIT, Transport and Energy:
The Challenge of Climate Change, Leipzig, 2008. Routier : 243 gCO₂/t-km;
Rail : 16 gCO₂/t-km ; Maritime : 19 gCO₂/t-km; Aérien : 1 054 gCO₂/t-km

Fig. 4.3 CO₂ emissions for freight transport modes. *Source* CAS [2]

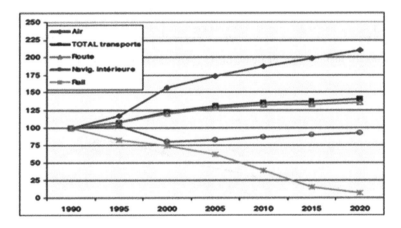

Fig. 4.4 Expected evolution of CO₂ emission by mode. *Source* Eurostat [17]

4.2.3.1 PRIMES Model

Rail freight energy consumption is also decreasing, compared to other transport modes (see Fig. 4.5). In 2006, 2.5 % of final energy consumption is accounted for by the transport sector in Europe a decrease of from 1990s 3.4 % [37].

Rail freight structural trend are thus lower as far as environmental externalities are concerned, but there is a strong correlation between this trend and the low rail freight activity and this is not a recent phenomenon. Figure 4.6 shows the decrease in the rail freight modal share compared to the other modes from 1970.

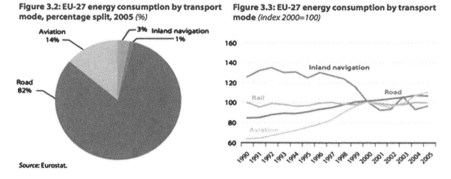

Fig. 4.5 Energy consumption in the European Union. *Source* Eurostat [17]

	Répartion modale des tonnes.km en Europe (%)			
	Véhicules Lourds	Fer	Voie d'eau	Oléoducs
1970	48.6	31.7	12.3	7.4
1980	57.4	24.9	9.8	7.9
1990	67.5	18.9	8.3	5.3
1994	71.7	14.9	7.7	5.6

Fig. 4.6 Modal split in freight transport: EU-15 (1970–1994). *Source* Crozet [13]

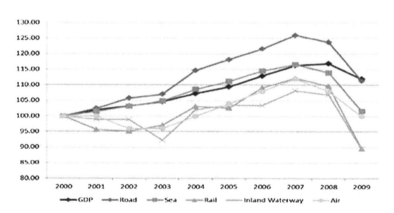

Source: Eurostat, International Transport Forum, Union Internationale des Chemins de Fer, National Statistics

Fig. 4.7 Modal split for freight transport (in value). *Source* Eurostat [16]

Figure 4.7 shows the recent decrease for rail freight transport (in value).

Rail freight tons-kilometres growth rate is lower than that of other transport modes. On average, the total transport annual growth rate is 2.7 % per year compared to 1.3 % per year for rail freight between 1999 and 2006 [8]. This trend is stronger than in 2008s economic crisis. Thus, EU-27 rail freight tons-kilometres decreased by more than 17 % from 2008 to 2009. They accounted for 366 billion tons-kilometres in 2009 [15]. In France, during the last quarter of 2013, only rail freight transport production decreased by a factor of less than 2.3 % [38]. This is even more worrying compared with the freight transport production's increase by 4.6 % for the same quarter. This increase is driven by waterway transport (+6.8 %), road transport (+6.5 %), air transport (+1.0 %) and maritime transport (+0.7 %).

There is therefore a gap between the increase in EU trade and the weak share of rail freight. Transport policies then focused on rail transport regarding environmental goals.

4.3 Sustainable Transport Policy and Innovation

In France, as in Europe, a sustainable transport policy is strongly needed because of the increase in environmental externalities (in terms of emissions and energy consumption). The sustainable dimension is explicitly identified in the European transport policy in 2001 [10]. However, since 1992, policies want to implicitly promote sustainable transportation. In this section, we provide a survey of evolution in policies towards sustainability.

4.3.1 Railway Packages

The first rail package represents a series of measures, introduced between 1991 and 2001, aiming at increasing rail's competitiveness, and including several liberalization-related goals. The initial goal consisted in separating infrastructure (network) management and transport operations. It then aimed at improving and securing infrastructure access for transport operators, by implementing a license system that is valid in all EU countries. Enabling fair competition between the various rail operators is another goal, at the same time guaranteeing financial sustainability of respective infrastructure managers.

In 2002, the EU edited a second railway package, completing the first one. The goal was then to increase security levels, and to enhance interoperability with regards to rail cargo market liberalization. It initiates the creation of national authority bodies, the development of common security methods in order to clarify respective responsibilities, the conciliation of security rules within the market liberalization context, as well as the improvement of transparency related to information exchange and accidents/incidents surveys. It also intends to impose a minimum service quality level for rail operators, to optimize cargo priority levels, to enhance rail transport's environmental performance at both emission and noise levels.

In 2004, the European Commission adopted a third rail package. It expressed a need to stress train conductors' training, by using a common approach for all member states. It proposed also to favor the (total) quality philosophy within the rail cargo industry, by offering adapted tools to railway operators. A fourth railway package is nowadays still under discussion, but we notice a growing awareness over the last few years with regards to social and ecological dimension's importance. The following section develops this increased awareness.

4.3.2 From Modal Shift to Co-modality

From 1992 to 2005, the European policy for sustainable transport of goods has been based on modal shift, which aims at transferring, for a constant traffic level, a part of traffics to modes with potentially lower societal effects. To goal was to transfer the growth of road traffic flows to more sustainable modes, like rail or inland water-ways. Europe wanted to overcome three major difficulties to lead transport to more sustainable solutions: the increasing inequalities between transport modes; the congestion of some major roads and railway lines; and negative effects on environment and citizens' health.

A break in policy happened in 2005 [26]. Modal shift has started to be criticized, notably because of its possible negative effect on economic growth. It's impossible to transport goods from door to door with modal shift. Offloading's take time and extra-costs thus reducing the competitiveness of modal shift and making it less profitable than road transport. In addition, the ASSESS study shows that the increase of modal shift would have negatives economic and environmental impacts. The mid-term white book of transports' assessment and globally speaking, the assessments produced during the last 15 years seem to contradict the allegations of positive effects due to modal shift. Rail transport has been strongly criticized by Vatanen during this transition from modal shift to "co-modality":

> Modal shift from road to rail is impossible and to try to do so would be contrary to the prosperity of Europe … To disconnect the increase of transport from economic growth is a good but wrong intention. Negative effects of transports have to be reduced, not transports. Many of their negative impacts are progressively disappearing thanks to more severe standards. Moreover, railways are far from armless from this point of view (…) Road brings more tax revenue than the investments they receive. It's the opposite for railways.

The European Commission then redefined sustainable transport using the notion of co-modality, which can be defined as "*the efficient use of isolated or combined modes of transport*" (CCE 2006). Today the goal is to optimize the use of transport networks. These networks can now be related to all the modes including road.

4.3.3 The Challenge of Intelligent Freight Transport Systems

To support co-modality, command of information plays an important role for European public policy.

In particular, improving the control of information is possible thanks to new management systems. Nowadays, Europe relies on intelligent freight transport systems. These systems would enhance a freight transport that meets the growing expectations in terms of information exchanges, of internationalization, security, tracking and paperless procedures.

The word "intelligent transport systems" refers to ICTs applications in transport. These technologies play a role to improve security and safety, to optimize the use of infrastructure, to limit energy consumption and emissions, and to favour the modal shift. ICTs represent 5 % of EU GDP, i.e. 660 billion euros, and employ more than 8 million people, i.e. 3.7 % of EU total employment (UE 2012). According to the European commission, this sector facilitates improved productivity in all sectors, including transport. Regarding transport alone, ICTs allow an increased mobility while reducing CO_2 emissions by 60 % by 2050.

Road mode is concerned with the deployment of the intelligent transport systems with: traffic management services; freight management services; telematics-based traffic information; and electronic toll system, but other modes are also concerned. For example, the European Rail Traffic Management System (ERTMS) whose main function is to follow and ensure a safe and sufficient spacing between trains. Air traffic as well as maritime traffic security will also benefit from this ICT development.

ICTs development aims at setting up integrated logistic and transport systems along with supply-chains. This integration should enable the combination of the specific qualities of each mode in order to provide shippers with better and more efficient service both economically and environmentally. *"Technological innovation is an excellent opportunity to integrate transport modes, to optimize their performance and improve their safety, and to enhance a more sustainable European transport system"* [10].

This ICT development to support sustainable transport can be identified at three different levels: international, European and national (basis on France in this chapter).

At the international level, technology is considered a significant means of sustainable development [32]. *"Technology and its transfer are central to current analysis of the relationship between economic development and the environment because they are the fundamental means for breaking the link between growth and pollution accumulation"* [29]. This technologies' impact remains strong in the more recent reports, in which technology is qualified as a key to favour more sustainable transportation [30].

At a European level, the first white paper of 1992 already argued for a technological cornerstone:

> Research and development action in the field of transport within the EEC should provide new tools.

In 2006 the fact is again underlined concerning the objectives of transport policy in terms of safety and international connectedness; innovation is key (CCE 2006):

> The policies of the EU aim to provide and commercialize tomorrow's innovative, energy-saving and alternative energy solutions and to support vast intelligent transport projects like Galileo.

One year later, the conclusions are the same [6]:

> Technology systems must enable transport to be environment-friendly, notably when it comes to greenhouse gas emissions.

In 2010, when the European budget was re-examined, the European Commission once again insisted on the importance of technological innovation.

> The Commission has put an increasing share of its policy, regulatory and financial levers at the disposal of Member States, regions and industry to foster investment in innovation. The Horizon 2020 Programme, in particular through its industrial leadership pillar, will provide close to EUR 80 billion for research and innovation. This includes support for key enabling technologies that will redefine global value chains, enhance resource efficiency and reshape the international division of labour. To facilitate the commercialization of research results, Horizon 2020 will also finance closer-to-market prototypes and demonstration projects than to date. A key element of the new Framework Programme is joining forces with the private sector through public-private partnerships in key industrial domains, so as to leverage further private investment.

Simulations on which the European Union is founded also emphasized the importance of using technological innovation to reduce greenhouse gas emissions [31]:

> In the specific context of European Transport Policy, this result has important consequences for subsidiarity issues. Furthermore, it is likely that an important contribution to the reduction of CO2 emissions will come from "emerging technology" instruments (with a large number of such instruments being described in further detail in the TRANSvisions Case studies). Given that new technology is invented and developed through the combination of a variety of factors, it can be seen that the implementation of technology instruments is not as straightforward (in a policy formulation sense) as the implementation of certain other types of instruments (such as road pricing or building new infrastructure). However, the EU can take a variety of actions to help the implementation of such instruments, where such actions can be classified under two general headings. Firstly the EU can provide financial support to help research and development of new technology. Secondly, once such technology is available, the EU can help its introduction through a variety of regulatory instruments and demonstration actions.

4.4 The Limits of Sustainable Transport Policy

The results of political efforts have been mixed concerning transportation compared with other sectors [40]. Several factors can explain these results.

4.4.1 Non Technological Innovation Less Encouraged

In the literature, innovation is often associated with technological innovation alone [35]. However, innovation is a multifaceted subject. As McCormick [25] indicated:

> The process of technological change is nothing less than a process of cultural transformation.

In terms of, non-technological innovation covers at least three dimensions: innovation supply, process innovation and business model innovation [28]. Innovation attempts to add value to a product, particularly in regards to the design or packaging. The process of innovation will often modify the organization, production or even the management of a company. Finally, the innovation business model changes the structures of company revenue. It covers the entire chain value of the business, such as the development of low-cost models. This typology provides an interesting synthesis, but without claim to completeness, although other types of innovation can be identified in the literature [20; 1].

Technological innovation's leadership in research works explains why public policies are nowadays more focused on technological innovations. Unfortunately, this leadership can have perverse effects. For example, the report from the economic council for sustainable development emphasizes "*the risk of focusing on technological innovations without distinguishing process, product and service innovations, as well as without considering their economic viability and the changes they imply. This logic leads to over-optimism in the emergence stage of new technologies and to severe corrections as soon as implementation difficulties appear. The internet bubble is an eloquent illustration*" [12]. Beyond the excessive reliance on the technological effects, focusing mainly on technological innovation can limit the positive effects of other types of innovation. So, some authors suggest the promotion of non-technological innovations to enhance new companies' practices [27]. Technological innovations are not always mutually compatible; these different technologies' effects can even be contradictory. A possible solution is to prioritize the different technologies, but the diversity of technological innovation as well as mass of associated data makes the innovations' ranking difficult [40].

Beyond these innovation specific factors, something's related to the transport sector require attention to the potential effects of technological innovation. On the one hand, the technology used in transportation is mainly derived from external innovation, for instance stemming from computer science or electronics. Transportation, as a sector, creates and circulates little specific innovation. However, the specificity of innovation in and by the transportation sector is more organization oriented: the command of long and complex logistics chains implies interesting organizational innovations. On the other hand, technological innovations in the transportation sector have reached maturity [40]. Incremental innovation is however weaker than in the nanotechnology sector. Consequently, what remains to increase transportation performance is radical innovation. Literature shows that this second type of innovation is more rare and more difficult to promote than the first [5], due to a technologically radical innovation needing to be supported by a specific organizational structure [39].

European transport policy seems to favour more and more innovation by exploration, i.e. looking for a suitable solution within a set of possible solutions. One of the shortcomings of this kind of learning approach is that the innovations of competitors are not known. The European Commission in [9] was made aware of the lack of coordination with respect to all possible innovations. Furthermore, the innovations are not well coordinated among different member countries, which diminished expected results:

> The investments indeed suffer from the defragmentation and inefficiencies due to different national programs.

Another form of learning is possible, Exploitation. Learning by exploitation allows concentration on one specific form of innovation and to optimize it [24].

In this section, we do not exclude potential gains from technological innovations. They can improve the efficiency of rail transportation. The reduction of noise, the improvement of brakes, and the increase of capacities are a number among the problems of which could be solved by new technologies. Railway coupling is one technique still to be developed in Europe. With automated couplers, coupling could be done without manual work between the railcars, which is both dangerous and takes time and is therefore costly. At the same time, it would allow an increase in capacity and the speed of the train without increasing the danger of derailment. Up to now, automated coupling is not done in Europe as opposed to the rest of the world [34].

4.4.2 The Neglected Service Innovation

Improvements occurred to develop a better knowledge about the different types of innovations. From the beginning of the 90s, the innovation survey from the Oslo Manuel has extended the innovation analysis. The same questionnaire is nowadays given to developed countries as well as emerging ones. At the European level, the Community Innovation Survey has developed in-depth and diversified methodologies for innovation analysis whose results enable us to have a better overview of the key drivers of firms' innovation. Without providing an exhaustive list, researches on these key drivers are numerous [33, 23, 22].

These surveys and research remain much focused on industrial activities, leaving the innovation analysis for services absent. However, logistic services have evolved a lot and this is not just an issue of handling and controlling goods. Nowadays, many different operations occur amongst them:

- Operations related to the transshipments, handling and controlling goods, but also storage, shelving processes;
- Operations linked to the final delivery: order picking, preparation of promotional consignment, bagging, price labeling;
- Management operations: order taking, tracking products expiry dates, inventory management;
- IT operations: inventory management, fleet management, order picking management, accounting, and data transmission.

Physical operations go along with IT operations to trigger other activities (digital transactions) on time while improving their monitoring (management IT). In addition, the integration of information systems enables the service providers to position themselves in noncore activities such as demand forecasting for suppliers, logistic design and engineering.

Services economics can take into account this complexity by dividing the final service in four operations [19, 20]:

- Logistic and material processing operations (M), involving the movement, transport, transformation and any other form of processing of tangible objects;
- Logistic operations and those involving the processing of codified information (I) (production, transformation, transfer, archiving and so on);
- Contractual or relational service operations (R) which consist of a direct service provided in contact with the customers and with a variable degree of interaction;
- Knowledge processing operations (C) using intangible technologies and codified methods and routines.

As far as innovation is concerned, research about effects of ICT in transportation confirm that this sector is constituted of firms providing mainly material operations, and others developing informational, relational and knowledge processing operations [14]. Thus, it is possible to highlight, from a dynamic perspective, several trajectories of service innovations in the transport and logistic sector. First, firms provided only material operations, then; through clients' expectations and suppliers' strategies they developed innovations to deal with information, relations and knowledge management.

The sustainability goal contributed also to this evolution in the service supply. As an example, we can discuss some typical changes in logistic and transport services towards sustainability:

- Informational operations enable the use of planning tools, such as ECR. The ECR tools promote an integration of logistic operations along the supply chain and a better bulking and so an optimization of the loading rate or even a decrease in the trips' number. This leads to an improvement of energy efficiency and to the development of long-term relationships between the supply chain's stakeholders (industrial firms and distributors);
- Development of tracking services, using ICTs and modifying informational operations, also highlights the growing awareness of sustainability in logistics.

Goods tracking contributes to meet the safety and security requirements, as well as to give greater transparency through information on the products' origin;

- Knowledge processing operations become more and more complex and implementing Supply Chain Management recommendations such as a better coordination between stakeholders and a real-time sharing of commercial information that is needed for the co-production of knowledge related to new tools or new management methods;
- Lastly, regarding relational operations, more and more ethics charters are signed between service providers and large retailers. These charters are intended to ensure compliance with rules regarding work conditions.

This trend towards sustainability can be found for all the operations that constitute the transport service on a broad and complex perspective including (on occasion) highly developed logistic operations. Hence, different trajectories of sustainability are developed according to the evolving requirements of the various operations within a single service. However, if the intelligent freight transport system can lead to changes in the services' supply, their influence on sustainable transport will be closely diverse whether they meet the shippers' expectations or not, as discussed in the following section.

4.4.3 The Neglected Shippers' Expectations

The firms' logistic and transport expectations are also strategic choices aiming at reconciling very different needs, i.e. constraints related to micro-economic time and costs optimization. Thus, according to the context, expectations will be different, as well as the used logistic and transport services. There isn't therefore a single type of logistic and transport strategy but diversity, related to the specific needs associated to specific contexts. That's why guidelines to promote sustainability will themselves be different. Four major options can be given as way of example. For some companies, sustainable transport means using non-road modes; for some others, it means optimizing the deliveries; still others choose to pool warehouses.

Therefore, the diversity of companies' logistic and transport strategic choices imply different effects of intelligent transport systems. An industrial logistic process, with stocks, needs basic services, with a lot of material operations. Evolution towards sustainability can in turn concern these material dimensions, and sensitivity to intelligent freight transport systems will decline. But the role of these intelligent freight transport systems grows, where the productive processes need more complex logistic processes. Material operations (including transport) become less important. Then, sustainability will be taken into account more in informational operations, relational operations and methodological or cognitive ones. The sensitivity to intelligent freight transport systems is also bigger (Blanquart et al. 2008).

Shippers have therefore different expectations that need to be taken into account to ensure new sustainable freight transport strategies. Some surveys exist in Europe to explore this diversity, as shown in Table 4.2.

Table 4.2 Motivation to integrate environmental innovations (the share per country can be read according to the number of innovative firms)

	Environmental regulations or pollution tax	Future environmental regulations or taxes	Public support or subsidy for environmental innovations	Customer demand for environmental innovations	Codes or agreements of good housekeeping practices
Belgique	20.1	16.3	7.6	13.6	26.1
Bulgarie	8.6	5.4	2.4	4.0	5.2
Rép Tchèque	40.6	26.8	9.5	13.6	24.3
Danemark	–	–	–	–	–
Allemagne	20.8	19.0	7.7	18.3	18.8
Estonie	24.1	19.3	4.4	17.2	26.3
Irlande	27.2	19.9	9.1	25.3	28.5
Grèce	–	–	–	–	–
Espagne	–	–	–	–	–
France	24.0	15.0	6.4	17.6	23.9
Italie	22.9	16.3	12.8	13.0	14.8
Chypre	7.2	5.3	3.1	3.9	13.1
Lettonie	19.1	11.3	8.3	13.6	34.0
Lituanie	39.3	31.8	12.5	26.8	24.5
Luxembourg	10.1	11.4	4.4	15.0	43.2
Hongrie	41.3	34.5	4.1	31.9	32.8
Malte	23.8	23.8	8.1	11.3	13.3
Pays-Bas	10.5	9.2	6.7	13.8	12.7
Autriche	–	–	–	–	–
Pologne	24.1	16.1	4.9	12.7	13.3

(continued)

Table 4.2 (continued)

	Environmental regulations or pollution tax	Future environmental regulations or taxes	Public support or subsidy for environmental innovations	Customer demand for environmental innovations	Codes or agreements of good housekeeping practices
Portugal	31.6	18.3	7.0	21.9	42.0
Roumanie	37.6	20.4	9.3	17.6	17.7
Slovénie	–	–	–	–	–
Slovaquie	37.0	27.3	4.7	11.7	18.9
Finlande	15.8	17.8	6.2	30.3	29.1
Suéde	8.4	12.3	2.7	14.7	15.1
Royaume-Uni	–	–	–	–	–
Croatie	35.7	28.0	8.4	19.6	30.3

Source Eurosat (code des données en ligne: inn_cis6_ecomot)

4.5 Conclusion

After examining the transport policy from 1991, transport policy has put more resources in one form of innovation to foster sustainability.

In the literature, innovation is often associated with technological innovation alone [35]. This technological innovation's leadership in research works can explain why public policies are nowadays mainly focused on technological innovations. Improvements occurred to develop a better knowledge about different types of innovations.

At the European level, the Community Innovation Survey has developed in-depth and diversified methodologies for innovation analysis, whose results enable us to have a better overview of the key drivers of firms' innovation. However, these surveys and research remain much focused on industrial activities, leaving the innovation analysis for services absent.

Finally, the firms' logistic and transport expectations are also strategic choices aimed at reconciling very different needs, i.e. constraints related to micro-economic time and costs optimization. Thus, according to this context, expectations will be different, as well as the logistic and transport services utilized. Therefore, there isn't a single type of logistic and transport strategy but a diverse array, related to the specific needs associated to their specific contexts, hence, guidelines to promote sustainability will themselves be different.

Therefore shippers have different expectations to be taken into account, to ensure new sustainable freight transport strategies.

References

1. Abernathy W, Clark KB (1985) Innovation: mapping the winds of creative destruction. Res Policy 14:3–22
2. CAS (2010) Le fret mondial et le changement climatique. Perspectives et marges de progrès, Rapport du groupe de travail présidé par Michel Savy. La documentation Française
3. CNT (2005) L'évolution sociale dans les transports terrestre, maritime et aérien en 2003-2004. Premier fascicule. L'évolution de la régulation sociale, La Documentation française
4. Citepa (2008) Air emissions-annual national data. Aailable www.citepa.org/emissions
5. Commissariat général au développement durable (2011) Cinq scénarios pour le fret et la logistique en 2040. Rapport PREDIT
6. Commission Européenne (2007) Plan d'action pour la logistique du transport de marchandises. UE, Bruxelles
7. Commission Européenne (2009) Un avenir durable pour les transports : vers un système intégré, convivial et fondé sur la technologie. UE, Bruxelles
8. Commission Européenne (2009) EU energy and transport in figures. Statistical Pocketbook. UE, Bruxelles
9. Commission Européenne (2013) Proposition de règlement du Conseil portant création de l'entreprise commune Shift2Rail. UE, Bruxelles
10. Commission Européenne (2001) Livre Blanc: La politique européenne des transports à l'horizon 2010: L'heure des choix. UE, Bruxelles

11. Commission Européenne (2006) Pour une Europe en mouvement—Mobilité durable pour notre continent. Examen à mi parcours du livre blanc sur les transports. UE, Bruxelles
12. Crifo P, Debonneuil M, Grandjean A (2009) Croissance verte. http://www.developpement-durable.gouv.fr/IMG/pdf/03-10.pdf
13. Crozet Y (2004) Les réformes ferroviaires européennes: à la recherche des bonnes pratiques, Institut de l'entreprise, coll. Les Notes de Benchmarking International, p. 93
14. Djellal F (2002) Innovation trajectories in the cleaning industry. New Tech Work Employ 17(2):119–131
15. Eurostat (2012) Decline in European road freight transport in 2011 reflecting the economic climate—Issue number 38/2012
16. Eurostat (ed) (2009) Panorama of transport. Eurostat Statistical Books
17. Eurostat (ed) (2013) Energy, Transport and Environment indicators. Eurostat Pocket Books
18. Gadrey J (1991) Le service n'est pas un produit: quelques implications pour l'analyse économique et pour la gestion. Revue politiques et management public 9(1):1–24
19. Gallouj F (1999) Les trajectoires de l'innovation dans les services: vers un enrichissement des taxonomies évolutionnistes. Économies et Sociétés, Série Économie et Gestion des Services 5 (1):143–169
20. Hollenstein H (1996) A composite indicator of a firm's innovativeness. An empirical analysis based on survey data for Swiss manufacturing. Res Policy 25(4):633–645
21. Mairesse J, Mohnen P (2005) The importance of R&D for innovation: a reassessment using French survey data. J Tech Trans 30(2_2):183–197
22. March J (1991) Exploration and exploitation in organizational learning. Organ Sci 10 (1):71–87
23. McCormick K (2002) Veblen and the new growth theory: community as the source of capital's productivity. Rev Soc Econ 60(2):263–277
24. Meunier C, Zeroual T, (2006) Transport durable et développement économique. http://% 20developpementdurable.revues.org/3305
25. Mongo M (2013) Les déterminants de l'innovation: une analyse comparative service/industrie à partir des formes d'innovation développées. Revue d'économie industrielle 143(3):77–108
26. Morand P, Manceau D, (2008) Pour une nouvelle vision de l'innovation. In: La Documentation Française (ed), Paris
27. OCDE (ed) (1997) Les instruments économiques des politiques d'environnement en Chine et dans les pays de l'OCDE, Paris
28. OCDE (2010) Perspectives des transports 2010. Le potentiel de l'innovation, Paris
29. Petersen MS et al (2009) Report on Transport Scenarios with a 20 and 40 year Horizon. Final report,Copenhagen
30. Prades J et al. (2005) Vers une stratégie de transport durable fondée sur le développement de l'innovation technologique. Esprit, Critique vol 7
31. Raymond PA et al. (2007) Flux and age of dissolved organic carbon exported to the Arctic Ocean: A carbon isotopic study of the five largest arctic rivers. Global Biogeochem Cycles, 21, GB4011. doi:10.1029/2007GB002934
32. SETRA (2013) Fret ferroviaire: quelles évolutions technologiques pour les wagons de marchandises. Revue Transports 25:10–16
33. Salter A, Tether B (2006) Innovation in Services: through the innovation glass of innovation studies. Paper presented in Background paper for Advanced Institute of Management (AIM) Research's Grand Challenge on Service Science, Tanaka Business School, London, 7 April 2006
34. Savy M (2009) Questions clefs pour le transport en Europe. La Documentation Française, Paris
35. Savy M (2010) Le fret mondial et le changement climatique, Collection Rapports et Documents, Paris, Centre d'Analyse Stratégique
36. Soe S (2013) Chiffres clés des energies renouvelables. Paris: SOeS, Ministère de l'Ecologie, de l'Energie, du Développement durable et de l'énergie. http: http://www.statistiques.

developpement-durable.gouv.fr/fileadmin/documents/Produits_editoriaux/Publications/Reperes/2013/reperes-energies-renouvelables-juin-2013.pdf

37. Steinmueller WE (2000) Will new information and communication technologies improve the 'codification' of knowledge? Ind Corp Change 9(2):361–376

38. Theys J (2007) Quelles technologies futures pour les transports en Europe?. In: CPVS (ed) Contribution au groupe—Technologies clés—de la Commission européenne

Chapter 5
Maritime Ports and Inland Interconnections: A Transactional Analysis of Container Barge Transport in France

Emeric Lendjel and Marianne Fischman

Abstract Recent research on maritime ports hinterlands points out the relevance of mass ground transport modes such as barge transport for enormous flows of containers to and from harbours, especially when a maritime port is located at the mouth of a river. Though, the modal share of container barge transport in French maritime ports is significantly lower than elsewhere. Some reports and studies explain the viscosity of container barge transport flows as a result of several factors, generally concentrated around the seaport community. In continuation of previous seminal works, this paper adopts a neo-institutional approach (Williamson in The Economic Institutions of Capitalism. The Free Press, New York, [51], The Mechanisms of Governance. Oxford University Press, Oxford, [52]) of container barge transport to understand how the factors generating this viscosity are managed. Section 5.2 describes the characteristics of the transaction of container barge transport. Section 5.3 is devoted to its attributes (asset specificity, frequency, uncertainty). According to Williamson's (The Mechanisms of Governance. Oxford University Press, Oxford, [52]) remediableness criterion, the observed governance structure of a given transaction is presumed efficient and aligned to its attributes. Thus, Sect. 5.4 deals with observed governance structures of container barge transport chains with a focus on Le Havre, main French container seaport and shows how agents try to limit opportunism in ex-post haggling over quasi-rents or under-investments. Implementation of a new institutional environment to modify governance structures is analysed, and a comparison with currently implemented governance structures observed in Rhine is made. Finally, Sect. 5.5 suggests ways of dealing with the remaining coordination problems impeding the development of container barge transport in France.

E. Lendjel (✉) · M. Fischman
Centre d'Economie de la Sorbonne (UMR 8174 du NRS),
Université Paris 1 Panthéon-Sorbonne, Paris cedex 13, France
e-mail: lendjel@univ-paris1.fr

© The Author(s) 2014   67
A. Hyard, *Non-technological Innovations for Sustainable Transport*,
SpringerBriefs in Applied Sciences and Technology, DOI 10.1007/978-3-319-09791-6_5

Keywords Container barge transport · Transactional chain · Transaction cost economics · Interface seaport · Governance structure

5.1 Introduction

What's the problem with barge transport in French seaports? Recent research on maritime ports hinterlands [21, 34, 38, 41, 43] points out the relevance of mass ground transport modes such as barge transport for enormous flows of containers to and from harbours, especially when a maritime port is located at the mouth of a river.[1] Though, the modal share of container barge transport (CBT) in French maritime ports (9 % of TEU in Le Havre and 5 % in Marseille in 2007) is significantly lower than elsewhere (32 % in Rotterdam and 33 % in Antwerp) [9, p. 46].

Research and studies that have been performed [4, 8, 17, 21, 22, 26, 45] indicate that the viscosity of CBT flows in France arises from several factors, generally concentrated around the seaport community. It may thus be noted:

- The lack of maintenance and investment in infrastructures [4, p. 61] which leads to the classic vicious circle (depicted in a cobweb diagram in [3, p. 327]) in public transports: if a public decision maker invests according to the observed traffic, a minority mode such as inland waterway transportation will obtain insufficient investments, generating a progressive degradation of its transport conditions and, thus, a decrease of its traffic, etc. From this point of view, traffic and volumes handled in main French inland ports are stagnating (none of them have reached their 1980s level) [47]. Except for containers, current handled tonnages do not motivate the public decision maker to invest in inland waterway infrastructures [8, p. 58].
- There are high transhipment costs in waterway transport compared to trucking [8, 17]. For (1) the quantity of transhipments is significantly higher in CBT. Usually, cartage (yard moves) is needed to move containers from the maritime quay to the river wharf, adding at least two movements of containers compared to trucking. As with any intermodal transport, CBT involves pre- and end-haulage carried out by road in order to move the container from its loading point to its unloading terminal. Hence, the total amount of transhipments is almost twofold compared to container trucking. (2) On French maritime ports, gantry cranes are used, both for barges and vessels, and are sized for the largest containerships. Using these cranes is expensive for barge transport operators, much more than the usual handling costs supported in trucking or even in rail freight. Moreover, scope and scale economies reached in Rotterdam or Antwerp are limited in French ports due to their lower overall performance in relation with other competing ports of the same range [40]. Since a lower volume of

---

[1] Let's remind that a Jowi class container barge can carry up to 500 TEUs per trip [6], 250 times more than one truck.

containers is handled, economies of scales are lower than at Rotterdam and Antwerp. Thus, the average unit cost of handling is higher in French ports and impacts the cost of containers loading and unloading on barges [35]. (3) The use of dockers—with specific status—for transhipment in maritime ports raises additional costs for barge transport operators compared to truck or rail handling. All these costs reduce significantly the a priori comparative advantage of waterway transport. Even in Rotterdam, 'container handling (move) is about 30 % more costly for a barge than for a truck' (*idem*. p. 42).

- The productivity levels of other assets of the CBT chain affect also its competitiveness and attractiveness. Note particularly that barges are subjected to unproductive waiting time for quay access, or before handling operations start and, of course, during loading and unloading… [1, p. 97]. For instance, as a delayed ship is more expensive than a barge, the docker staff is assigned to the ship, even if it means a waiting time for the barge [37, p. 291].

This list is not meant to be exhaustive. It just points out several factors affecting the transaction of CBT in France. In continuation of previous works [10, 18, 19, 44, 49], a neo-institutional approach [51, 52] of CBT was chosen to understand how these factors are managed. Transaction cost economics is a useful tool to tackle with coordination problems and to explain actual governance structures (market, hybrid, hierarchy) and strategies of firms to control this transaction chain. Few studies address the specific coordination problems affecting CBT in French seaports, despite their influence in shippers' modal choices [23]. We suggest that these difficulties could come from an insufficient degree of integration of this chain.

Thus, next section (Sect. 5.2) starts with the characteristics (nature, perimeter, main features) of the transaction of CBT. The third section (Sect. 5.3) is devoted to its attributes (asset specificity, frequency, uncertainty). According to Williamson [52] remediableness criterion, the observed governance structure of a given transaction is presumed efficient and aligned to its attributes. Thus, the fourth section (Sect. 5.4) deals with observed governance structures of CBT chains in Le Havre, the main French container seaport. We point out that the need for control explains the features of the observed structures on the Seine river. The last section discusses (Sect. 5.5) the nature of remaining coordination problems impeding the development of CBT in France.

5.2 Definition and Perimeter of the CBT Transaction Chain

According to Williamson, 'transaction occurs when a good or service is transferred across a technologically separable interface' [52, p. 58]. Here, the transaction at stake (CBT) is to transfer a service of transport, so that a container can be moved from one port to another during a given time. This apparent simplicity implies many sub-transactions, slightly different from those in bulk transportation [18] yet involving nearly the same actors.

A container is designed to carry quite any kind of cargo that can be unitized and to be used recurrently. Hence, goods are not moved anymore, but their containers. By dissociating the vector of transport and its container, this modality of inland shipping adds up a sub-transaction—the supply of container—compared to bulk transportation. But even if a container is an essential asset for the CBT transaction, it remains peripheral as a maritime loading unit. CBT can, therefore, be considered as a sub-transaction of the container transport's transaction. Indeed, most often maritime transportation is the main link of this transactional chain, and inland waterway transport is but the hinterland transport of shipping lines. It usually is the forwarder or the shipping company who has the commercial function to fill containers and sometimes to organise their packing/unpacking.[2] In other words, container's commercialisation (and thus its management) is usually peripheral to CBT.[3] A specific extraction from the ECHO[4] national survey data base (realized by the INRETS in 2004) shows that 100 % of the 23 CBT shipments found in the survey (from a total of 10,462 shipments involved in the survey) are outsourced and involved at least three operators. Hence, the CBT is a link in a larger chain and usually more complex to organize than the other transport chains, as only 7 % of the shipments—parcels excluded—involve at least three operators [5, 28, p. 108, 8].

Above all, it is necessary to distinguish these sub-transactions and physical flows. At least three flows can be involved in this chain, implying a complex coordination between them: the container's flow is potentially the most complex, since it goes from the sender to the recipient and involves non-barge transport links; the barge's flow does not only link a port to another but, it, also, moves inside a harbour from one quay to another; the power pusher-tug's flow (with a high power of about 2,000 cv) moves a convoy of barges lashed together to link two ports; the harbour pusher-tug's flow (with a low power of about 800 cv) deals with the positioning of barges in a port from one quay to another one (Fig. 5.1).

Managing the coordination of these flows has consequences on the governance structures' choices, as it will be explained in Sect. 5.4. For the moment, this hierarchy helps to restrict the perimeter of the CBT transaction. It includes at least six sub-transactions (ST).

1. ST 1 is the transfer of organisation and coordination of a container transport by the shipper to an economic unit (usually different from the shipper, corresponding to the missions of a freight forwarder, but can be performed by a shipping company).

---

[2] In most of the case, the shipper fills the container in his warehouse.

[3] Note that it could be part of the transaction's perimeter when CBT become independent of the maritime chain as it is on the Rhine [55].

[4] ECHO ("Envois-CHargeurs-Opérateurs de transport") is a national survey designed to understand shipper's practices and who's measurement unit is the shipment sent by a shipper. For a summary of the ECHO survey's results, see [28]. The very few observed shipments (70) involving a barge transport in the survey do not permit a quantitative analysis. Hence, our analysis of the data is "qualitative" and the proportions given cannot be statistically representative in the insight of the small sample size. Yet, they are not irrelevant.

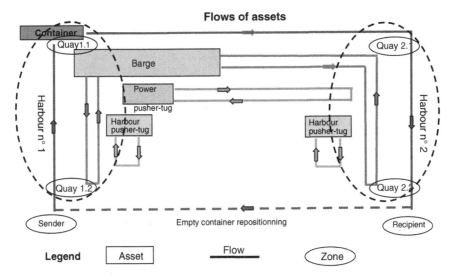

Fig. 5.1 Flows of assets

2. ST 2 is the transfer of the rights to use a transport capacity between the transport organiser and the owner of barges (or of slots) or of containers.
3. ST 3 is the transfer of the CBT service to provide quay to quay, from the transport organiser to the CBT economic unit. This sub-transaction is itself likely to be split in two: the CBT itself and the barge propulsion by a pusher-tug and a crew
4. ST 4 is the transfer of the service of cargo loading on barges between the transport organiser (often the shipping company) and the handling company at the departure quay.
5. ST 5 is the transfer of the service of cargo unloading from barges between the transport organiser (often the shipping company) and the arrival quay handling company.
6. ST 6 is the transfer of the service of ports interconnection (signalisation, locks, dredging, etc.) between a river infrastructure operator and a user.

Each of these sub-transactions includes many sub-sub-transactions. Handling operations (loading/unloading containers, managing containers at quaysides) are in themselves a complex ST set, especially since the container revolution. Indeed, the terminal operator must be able to implement the vessel-loading plan sent by the shipping line operator. Thus, he has to face several external constraints to comply with container handling procedures, with a precise order of loading/unloading containers and, therefore, with onshore containers pre-arrangements [54]. He also faces his own constraints (minimal number of container movements, optimisation of available space, management of human and material resources, and so on) to comply with the plans [25]. Though interdependent, the container loading/unloading ST is

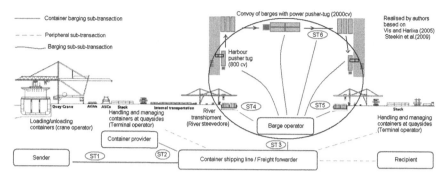

Fig. 5.2 CBT transaction

sometimes dissociated from quay container handling ST. As for container inland navigation operation itself, it is likely split into the CBT operation and the barge propulsion by a pusher-tug, the latter being sometime divided in two parts: the availability of a crew and the provisions of a pusher-tug. Other STs may occur when local pushers (or harbour pusher-tug) move barges in a port, whereas power pusher-tugs make convoys of lashed barges for long hauls between ports. Thus, the apparent simplicity of the CBT transaction is actually more complex, as illustrated in Fig. 5.2.

According to a well-known mechanism, expanding demand by container standardization leads to the segmentation of the barge transport transaction into an increasing number of sub-transactions. Nevertheless, as one will see in Sect. 5.4, the actual number of governance structures framing these sub-transactions is surprisingly reduced compared to what is theoretically possible. The study of the transaction attributes is thus required to explain it.

5.3 The Transaction Chain Attributes

Following [52], a transaction is characterized by its attributes: asset specificity (i.e., level of loss associated to alternate use of assets involved in the transaction, or redeployability's level of the asset), frequency (number of times a transaction occur in a given period), uncertainty (related to the environment of the transaction and to the behaviour of parties in condition of bilateral dependency). Only those which affect the efficiency of the CBT transaction chain, particularly in port connection, will be analysed here.

5.3.1 Assets Specificity

Williamson [52, p. 59] recognises six kinds of asset specificity: site specificity, physical asset specificity, human asset specificity, dedicated asset, reputation asset, temporal specificity.

Specificity of site The first kind of asset specificity used by Williamson is

site specificity, as where successive stations are located in a cheek-by-jowl relation to each other so as to economize on inventory and transportation expenses (*idem*).

Maritime ports are highly site-specific for inland navigation agents [19]. Indeed, these ports are places of maritime container departures and arrivals. Inland quays must, thus, be located in the vicinity of maritime quays to minimize costs (in space and time) between these two modes of transport.

Indeed, except for direct transhipment from ships to barges (as for instance in Hong Kong port, [24]), handling of maritime containers required to move containers between inland and maritime quays (cf. Fig. 5.2). In some ports, such as Le Havre [20, p. 17], this interconnection is minimized because the maritime quay is also an inland quay. But this multifunctional quay is costly for barge transport operators (oversized equipment, random availability of the quay...) (*idem*, p. 27) and advantageous for the terminal operators (higher rate of handling capacity utilization). It is even more expensive at the new Port 2000 in Le Havre, since there isn't any direct access for barges. Only multipurpose vessels (ocean/river) can reach it (like those recently put into operation by two French barge operators, Logiseine or River Shuttle Containers), but with much higher cost. Barge operators are faced with the trade-offs between extra costs of maritime handling (but at the vicinity of inland quays) and additional costs of a dedicated handling quay with the related transfer of containers from inland to the maritime quays. The development of extended gate model [43] is certainly a way of de-specifying maritime port for shippers but not for barge operators.

This very high degree of site specificity requires landowners, such as national government, regional state or city[5] to deal with this issue [27]. The exceptionally long duration of contracts (from 15 to 30 years depending on terminal) of land concession (quasi-integration of site) is an indicator of the high level of site specificity and of the quasi-rent it generates [27, 39]. The length and the frequency of port strikes in France could be seen as a proxy of its significance for the port community and of tensions over its sharing, as De Langen noted it in 2007:

The presence of economic rents may partly explain the strong bargaining position of port labour and resulting wage level and labour conditions [11, p. 463].

In fact, a strike highlights the risk of hold-ups generated by this specificity of site. The legal monopoly of port labour, before the French port reform, empowers one of the actors in the struggle for this quasi-rent distribution.

Physical assets As for any network activity, transport infrastructures are an essential asset for transport service. Their features (draft and air draft, width, locks' capacity, restrictions of navigation during nights and week-ends, etc.; see [30,

---

[5] Since the law n° 2008-660 of July, 4, 2008, related to the port reform, "Great Maritime Ports" in France are owners of the state properties (Art. 15) except those which belong to the maritime public domain or to the natural waterway public domain.

pp. 30–32]) play an essential role in the transport service's efficiency. They constrain the production units—hence the functions of production—of barge operators.

Assets in large gauge basins have an average payload of 200 containers (TEU) on the Seine, and up to 500 containers for the biggest self-propelled barge, e.g. Jowi [6, p. 51]. These assets, with integrated propulsion and hulls specifically designed for containers, have a high level of specificity. Since there is a lack of large gauge interconnections between the main French inland waterway basins (Rhine, Rhône, Seine), barges on them have a high specificity. Their transfer to different basins requires a maritime move with a tugboat, implying high redeployment costs. By linking Seine and Scheldt (and thus Rhine) in 2017, the Seine-Nord Europe project will thus reduce the specificity of these assets, since 'Seine and Oise will be branches of the Rhine basin'.[6] From this point of view, barges and pushers have a lower specificity since they can be separated and connected quite easily even if they work together (see Fig. 5.2). They can thus easily be used for other transports than containers. They are standard assets. Their necessary interdependence gives them a low specificity.

Note that containers are highly standardized physical assets (and thus with a low specificity) but, as quoted before in Sect. 5.1, they are essential to transactions. Hence, they can have a higher specificity for barge operators who use them apart from the maritime chain since the activity of barge operators depends, thus, on the volume of maritime containers at their disposal [55, p. 43].

Human assets Container river transport transactions require, as in bulk, highly specific human skills both upstream (commercial skills, organizational skills, etc.) and downstream (related to transport service itself).

A bargeman, whether employee or not, accumulates knowledge—often informal—and skills about his boat, the river basin, the practices of loading/unloading in different ports [18]. Boatmen acquire a perfect knowledge of their navigation area and must rely on the knowledge of other boatmen when they venture on another waterway. When they do not own the boat, they accumulate experience and expertise on specific equipment (pushers and barges) they will not find elsewhere in another French company.

As regard the handling, the former legal monopoly of dock workers and cranes drivers (French law of September, 6, 1947) before the 1992 and 2008 reforms, gave them a high level of specificity since any handling operator had to employ them for maritime handling operations. It remains after the reform in a lesser extent due to the European principle that no ships can be "self-handled" [48]. Yet, loading and unloading barges may not be necessarily done by employees of maritime stevedoring companies (art. R. 511-2 of the French transports Code) if the barge operator owns a dedicated barge terminal. But, as a matter of fact, this situation remains at Le Havre.

---

[6] Interview of Mr. Fortrye (CFT), November 2010. Yet, remaining gauge differences between these two rivers (IAU 2008, p. 31) will prevent any Jowi class vessel to move on the Seine till Gennevilliers. It will always stay on the Rhine only.

This high specificity strengthened after the container revolution. The handling job has indeed changed due to the increasing technicality and capitalistic intensity of the operations needed by containers. Hence, the acquisition of more technical knowledge and a higher degree of dock workers and cranes drivers specificity. The intermittency associated with dock workers legal monopoly was justified by significant fluctuations of activity due to tramping [8, 25]. The container regular line development has deeply changed the frequency in handling and the degree of specificity of the assets used. A regular activity requires the regular attendance of qualified staff, which has also reinforced this high degree of specificity. Hence, a discrepancy arose between the governance structure (intermittency) based on an institutional environment and the attributes of the port handling transaction.

Dedicated assets Carrying containers entails dedicated assets from barge and terminal operators to reach high levels of productivity. Terminal operators need dedicated specific cranes and engines (appropriate berth, gantry cranes, stacking straddle carrier, stacking space,...) to secure container handling. Barge operators need specific barges designed to carry containers and dedicated terminals to load/unload them. The mutual dependency generates risks of under-investments and quasi-rent [32, 33]. This concern is particularly high during the launching phase of a new service, as was the case for CBT on Seine in 1994. A barge transport operator (Logiseine) and terminal operators in Le Havre and in Gennevilliers (near Paris) have had to generate simultaneously large investments to start a regular line. If maritime ports can easily allocate part of their resources to barge calls (when free of container ship to be loaded/unloaded), this is not the case for inland ports which first need to invest. The large amount of required capital and its negative profitability in the short term explains why public port authorities are often involved during the launching phase of a CBT line. For instance, Paris Terminal SA, the inland public port operator, was a stakeholder of Logiseine barge operator at its very beginning. Based on Joskow [32] and Klein [33] we know that vertical integration is likely to occur when dedicated assets are involved in the transaction, in order to avoid hold-up risks associated to the quasi-rent at stake in dedicated investments. Thus, unlike Franc and Van der Horst' Resource-Based View assertion, it is not "to convince shippers of the shipping lines ability to secure container flows, and consequently to offer reliable services" [19, p. 561] that operators integrate site assets, but because these assets are dedicated assets and thus generate interdependency.

Time assets In order to minimize loading and unloading times, terminal operators identify time slots to allocate to their clients. These time windows help them to optimize their handling means and container barge operators to minimize their waiting times. Oddly, if these assets are the primary purpose of the agreements between terminal operators and barge operators on the Rhine [55, p. 29], they are rarely subjected to recoverable property rights (unlike air slots in the United-States air transport system, for instance) though quasi-rent involved is far from being negligible. It is the same for container storage time at terminals, even if it is now frequently subjected to a particular pricing to avoid congestion [43, p. 175]. This (and also lock time slots) is crucial for agents' coordination problems in ports, as we will see it in Sect. 5.4.

5.3.2 *Frequency and Uncertainty*

Container standardisation lowering transaction costs promotes the positive effects of the market governance structure, i.e. scale and scope economies due to the aggregation of diverse demands [51, 52, p. 92, 66]. Indeed, the unit cost of transport of a container will be less for a shipper if fixed costs of a boat with its crew are divided between several shippers. Transacting with a barge operator is less expensive here than the ownership of a dedicated fleet by each shipper. Thus, whatever the frequency of transactions is, a shipper has no incentive to internalize CBT owed to administration costs it generates and the loss of scale and scope economies reached by the market.

But it is different within the transactional chain. The frequency and the regularity of transactions between each link in the chain is high enough to justify a more integrated governance structure, particularly when assets are dedicated to the transaction. Frequency is the main (but not the only) factor justifying integration (or quasi-integration) of the chain. This is the case of the transport capacity needed to reach the frequency of the regular line service provided. For instance, with four shuttles per week between Le Havre and Gennevilliers [30, 50] Logiseine can't rely on spot transactions to purchase every week the human and physical assets (with more or less high degree of specificity) needed, except during peak activity to increase its capacity. It is the same for all the other operators due to the networked nature of this activity.

This logic also prevails with propulsion and support sub-transactions in the CBT transaction. Integration of propulsion can technically be made on a self-propelled boat, so that propulsion costs decrease due to a better hydrodynamic shape of the hull. It can also be done by keeping separate the elements of propulsion (pusher) and the elements of support (barges). Integration is not technical then but only transactional. For barge operators, this dissociation has the advantage of avoiding waiting time for loading and unloading goods as barges only stay at quays.[7] It could also help the operators of a waterway basin to optimise the number of barges needed for the volume of transactions [55, p. 17].

Likewise, a high and regular frequency of handling transactions can be observed in ports too. Starting from Le Havre, Voies Navigables de France (VNF, the French transport infrastructure manager) indicates an average of three services per day (from Monday to Friday), with one or two more departures if necessary (VNF 2009, p. 33). Arrivals are significantly less distributed with two activity peaks (5 services) on Monday and on Wednesday morning, the other days of the week having one or two services only. Given the high degree of specificity of physical and human assets in handling (and of scale economies they gain with the recurrence of transactions), it is not surprising that these assets be, as for quays, integrated by stevedoring companies.

---

[7] These waiting times always have a random dimension since a third party can always influence the transaction (externality) and affect then the possible frequency of the transaction [18].

The last factor affecting barge transport transactions is uncertainty. Williamson [51] distinguishes two sources of uncertainty: behavioural uncertainty (coming either from the bounded rationality of agents or from opportunism) and environmental or institutional uncertainty (change in demand, technical progress, change in regulation, etc.) [16, 36]. Environmental uncertainty is prevalent in CBT [19], especially in France, because of the new Seine-Nord Europe channel which should be open in 2017 [2]. The opening of the Seine basin is likely to change the governance structures because of the increased traffic (as we will explain it below).

5.4 Which Alignment of Governance Structures to the Transactional Chain Attributes?

As Williamson stated:

the critical dimensions for describing alternative modes of governance (...) are incentive intensity (...), administrative command and control (...), and contract law regime (...) [53, p. 681].

Following this typology, structures governing first and second level sub-transactions in France in waterway regular lines have now to be assessed.

5.4.1 First Level Sub-transactions in France

The first point to consider is the theoretical number of combinations of governance structures permitted by the transactional chain of the river transport of containers. Considering that each of the five sub-transactions (if the river interconnection service is excluded as public service cf. Sect. 5.1) may be achieved by at least three governance structures (spot market, hybrid or long-term contract, hierarchy), the decision tree includes a set of 243 possible combinations ($3^5 = 243$) of governance structures for this transactional chain [53]. The longer the chain, the bigger the set of possible combinations.[8] This exponential character of the economic complexity of the chain contrasts sharply with the very limited number of combinations observed in the river transport of containers in France. If an inventory of these chains has already been done [21, 30, 55], VNF 2009), their typology and their understanding in the insights of transactional analysis has still to be done for the French basins.

---

[8] Be n the number of elements in the subset of governance structures and p the number of transactions at stake, n^p is the total number of available combinations. Thus, if all the governance structures are taken into account (franchise, joint-venture, quasi-integration, long-term agreements, etc.), the field of potential increased more quickly.

Among the six services observed on the Seine river between Le Havre and Paris[9] (Fluviofeeder, Logiseine, Maersk, MSC, RSC, SNTC Carline), four services are really provided by barge operators, Maersk and MSC having slots on Logiseine's barges (VNF 2009), acting as shipper for Logiseine (see Sect. 5.2). All of the governance structures of those CBT operators on the Seine are integrated or quasi-integrated:

- Fluviofeeder is a subsidiary of Marfret shipping company. Through a partnership with MSC, Fluviofeeder provides a regular line between Le Havre and Rouen. Here, the vertical integration involves forwarding and CBT quay to quay only.

- Logiseine, CBT operator, is a 'Société en nom collectif' associating terminal operator companies from the two ends of the chain (Terminal of Normandie, in Le Havre, and Paris Terminal S.A., in the Port Autonome of Paris-Gennevilliers) and the barge operator, CFT. The transaction chain integration is complete here, from forwarder to handling operations. The withdrawal of PTSA from this joint venture in December 2006 doesn't mean a de-integration since TN and CFT are still owners of PTSA [30, p. 44]. Henceforth, CFT owns 55 % of Logiseine capital and TN 45 %. Note that Logiseine owns floating cranes to handle its own operations if necessary and the operator owns 15 % of the shares of the "Société d'Aménagement de l'Interface Terrestre du havre" (SAITH) which runs the railway line between maritime terminal and quay of Europe dedicated inland terminals.

- Maersk shipping line provides a carrier haulage service on the Seine river in the continuity of its maritime service. This is an example of vertical quasi-integration by a ship owner of the river link since Maersk charters transport capacity (barges and slots) to Logiseine with long-term charter contracts. Through Logiseine, Maersk also controls the barge handling in Le Havre and container's cartage.

- MSC shipping line has a similar strategy as Maersk. It charters transport capacity to Logiseine. But the partnership with Terminal of Normandie to operate on Bougainville terminal allows MSC to control barge handling more directly (but still hybrid) than Maersk does since this terminal is dedicated to barges too.

- The vertical integration is even deeper for the owner CMA-CGM since it not only owns the barge operator River Shuttle Containers (RSC), but also 35 % of the capital of the SAITH's railways line via the Générale de manutention Portuaire (GMP) in Le Havre, a joint venture of CMA-CGM with DP World. GMP operates the dedicated barging terminal of Europe's quay.

- SNTC-CARLine operator is particularly interesting in the insight of neo-institutionalism since this company was created from a partnership between a shipper/ freight forwarder specialized in grains, Soufflet, an inland navigation cooperative, the SCAT, and a truck operator, STTI, member of a group of carriers ASTRE. Here is a form of horizontal integration between small or medium size companies to provide an integrated offer of carrier haulage services (see Table 5.1).

---

[9] The services observed in Dunkirk and Marseille are similar, so they will not be described here. See Frémont et al. (2008) for instance.

When a barge operator has no specific agreement with a handling operator, its handling agreement is usually settled on an annual basis even if it is proportional to the volume of containers handled. Thus, strictly speaking, there isn't any arm's length transaction. Hence, this table shows a systematic quasi-integration of the transaction chain by these six operators, with five different modalities. It is thus clear that among the three dimensions described by Williamson (incentive intensity, administrative command and control, contract law regime), control strongly dominates here. Hybrid modalities predominate when complete integration of the considered sub-transaction is not possible (lack of funds) or not wanted (for incentive reasons). As described in Sect. 5.2, the limited numbers of governance structures observed here can be explained by the attributes of the sub-transactions of that transaction chain. High frequency (and besides regular) of transactions and the (quite high) degree of assets specificity may also explain that four barge operators only can be found in CBT on the Seine river compared to almost 800 barge operators in bulk river transport [47].

5.4.2 Second Level Sub-transactions

Integration or quasi-integration also characterises second level sub-transactions (or sub-sub-transactions). It is not possible to describe here all the chains. But some phenomena are interesting in the light of transaction cost analysis. Let's note two contradictory movements in second rank sub-transactions. The sub-transactions in handling service tend to be integrated (following the trend observed in other big ports in the world) but those in barge transport itself follow a segmentation process and an increasing outsourcing trend.

In the sub-transactions in handling service, for instance, site specificity is so high in Le Havre that the railway line (operated by SAITH) between port 2000 and the dedicated inland terminal of Europe's quay involve almost all the main stakeholders (CMA-CGM via GMP which control also Europe's quay, Logiseine, indirectly MSC via TN's partnership) in its capital. The amount of dedicated investments and of flows to generate explains this quasi-integration. The natural monopoly, which characterized this rail service, requires its control by main stakeholders. Hence, small operators, like SNTC, can only protest against the prohibitive cost of this transfer (VNF 2009). Integration to control human assets (dockers first, then cranes drivers) has been done while market (via the dockers' intermittence system) prevailed before the 1992 and 2008 the docker status reforms.[10] It aims to integrate dockers in handling operator companies and thus to promote hierarchical coordination activities [25, 29]. The lobbying of the handling firms syndicate (UNIM) to change their institutional environment helped them to align governance structure

---

[10] Law n° 92-496 of June 9, 1992 changing the work arrangements in maritime ports and Law n° 2008-660 of July 4, 2008 related to port reform.

Table 5.1 Governance structures (M—Market, X—Hybrid, H—Hierarchy) on the Seine river

N° Transaction Label	Synthesis of GS for the transaction chain p	ST1	ST2	ST3	ST4	ST5	ST6
		Transfer of the container transport organisation between a shipper and a forwarder	Transfer of rights to use a transport capacity between the transport organiser and the owner of capacity	Transfer of the quay to quay transport operation between the transport organiser and the barge operator	Transfer of the maritime port handling from the transport organiser to the handling company at the departure quay	Transfer of inland port handling from the transport organiser to the handling company at the arrival quay	Interconnection
Fluviofeeder	XHHMM	X (Marfret's subsidiary; long termcontract with MSC) or M	H	H	M	M	M
Logiseine	MHXXX	M	H (Barges of Logiseine)	X (Logiseine is a Joint venture hold by CFT and TN)	X (Logiseine is a Joint venture hold by CFT and TN)	X (CFT and TN shareholders of PTSA)	M
Maersk	XXXXX	X (long-term agreement with CFT)	X (charter capacity to CFT)	X (Logiseine partially hold by CFT)	X (Logiseine is a Joint venture hold by CFT and TN)	X (CFT and TN shareholders of PTSA)	M
MSC	XXXXX	X (long-term agreement with CFT)	X (charter capacity to CFT)	X(Logiseine partially hold by CFT)	X (TN MSC is a joint venture with TN)	X (CFT and TN shareholders of PTSA)	M
RSC	XHHXM	X (CMA-CGM's subsidiary) or M	H	H	X (GMP is a joint venture hold by CMA-CGM and DPW)	M	M

(continued)

Table 5.1 (continued)

N° Transaction		ST1	ST2	ST3	ST4	ST5	ST6
SNTC-Carline	XXXMH	X (Alliance with Soufflet) or M	X (Alliance with SCAT)	X (Alliance with SCAT)	M	H (SNTC owns handling of Nogent/Marne)	M

and the container handling transaction attributes. Henceforth, handling operators have integrated cranes drivers and gantry workers in almost all the main maritime French ports [45, p. 26]. But the theoretical advantages of the intermittency were dockers' flexibility and redeployment as it still can be seen in Antwerp today.

Regarding transport, an opposite movement can be observed. As control of specific human and physical assets is needed, CFT has no other choice but integration or quasi-integration. Meanwhile, CFT wants to have a more incentive device than hierarchy. Thus a singular hybrid structure—a *Société En Participation* (SEP)—is sometimes used for some of its pushers.[11] The SEP pools resources from each partner. CFT rents the physical asset (the pusher) and a small dedicated company (SARL) rents human resources (two crews of six people to secure a 24/24 driving each week). The SEP sells a pushing service per hour or kilometre (with a yearly contract) to CFT. Logiseine commercialises the capacity of transport on its barges and asks CFT to make the transport service. Thus, CFT sub-contracts the pushing service to the SEP. The SEP assumes earnings and operating expenses of the pusher so that each partner is directly involve in its operating income. Particularly, the structure incites the crew to take care of the equipment (a pusher is very expensive, which explains why CFT needs to be part of the SEP), its fuel consumption (main variable cost) and the service liability and punctuality. Nevertheless this hybrid governance structure (i.e. quasi-integration) is often used in road haulage [15] but relatively uncommon in river transport, even with CFT. In contrast, other more usual hybrid structures (regular sub-contracting, long term charter, barge transport pool, etc.) can be seen on the Rhine [56].

5.5 Discussion of Remaining Coordination Issues

All the arrangements set up to align governance structures and the transaction attributes are not always enough. As said in Sect. 5.2, transaction needs to coordinate flows through a governance structure. But (1) voyage charter contract usually used in CBT involves agents only in a very limited part of the transactional chain; (2) externalities (due to third parties to the contract) affect barge operators.

5.6 Voyage Charter in CBT Transaction

CBT transaction transfers a service of transport between a shipper and a barge operator, so that a container can be moved from one port to another during a given time. Its governance structure is usually the spot market[12] (voyage charter contract).

---

[11] Interview of Mr. Fortrye (CFT).

[12] Interview of C. Rose, General Secretary of the French Association of Shippers (AUTF).

It is a transfer of rights (and thus of liability) on the container transport. This (limited) transfer to the barge operator starts *only when containers are loaded* (under the responsibility of the shipper) *on a barge*. It terminates once unloading is achieved.

Voyage charter is a governance structure, which includes

- the sub-transaction 1 of the CBT transaction;
- the sub-transaction 3 (transfer of the CBT service to provide quay to quay).

Hence, some essential sub-transactions *are out of a voyage charter*:

- the availability of a transport capacity (ST2), necessary for the barge operator to honour his contract;
- the handling service (ST4 and ST5), even if the shipper has to load/unload the container;
- the service of providing port and transport's infrastructures (the web, clocks, quays, etc.) (ST6).

Sub-transactions out of a voyage charter have other governance structures. ST2 usually has either hierarchy or hybrid (long term agreement) governance structures. ST4 and ST5 are governed by handling spot contracts, even if these last are also supervised by hybrid governance structures. ST6 occurs on each move, but through fees and taxes.

Thus, CBT involves several contracts, with a theoretically full mapping of liabilities. In case of problem occurring during sub-transaction ST2, ST4, ST5 or ST6, out of a charter voyage, (externality) the court procedure is complex and costly (since it implies a third party).

Thus, two differences could help to understand the low modal share of barge transport in French port hinterlands. Contrary to the Rhine River, (1) ST 4 and 5 are not achieved by maritime terminal operators and (2) terminal operators in Le Havre, Marseille, Dunkirk, Paris, Lyon or Lille are not contractually or commercially directly involved in the CBT transaction. Seaport stevedore companies are economically involved via their capital equity in CBT companies in France as on the Lower Rhine, but the lack of joint commercial structure, of the Extended Gate model kind [43], drastically reduces their convergence of interests with other river transport agents to develop this mode of freight transportation.

5.7 Externalities and Voyage Charter

Externalities undergone by barge operators (In Le Havre as any other maritime ports) are mainly port congestion and waiting time [17, p. 26]. These externalities have significant consequences on the regularity and the reliability of container river transportation. According to the Cour des Comptes [9, p. 109], waiting time for access to terminals, cranes failures and strikes cost 500,000 € to Logiseine in 2004. For instance, access priority to vessels instead of barges is very costly for Barge operators. Far from being insignificant in shipping, vessel delays concern 40 % of them [46], upsetting the Handling operator's planning (and hence of barge operators). Their significance in Antwerp and Rotterdam has generated several research papers [7, 14].

As well known, 'a primary function of property rights is that of guiding incentives to achieve a greater internalization of externalities' [13, p. 348]. Externalities suffered by barge operators exist only because of very significant transaction costs to define and allocate property rights. The question is whether it is possible (and at which amount of organizational cost) to define property rights on temporal assets (see Sect. 5.3.). The shadow price method is the usual approach to determine the amount that operators will accept to pay to relax or dissolve this externality [31, p. 12], the cost to get rid of it. In CBT, valuing waiting time is possible but complex since the delays of the barge and its pusher (measured in hourly cost of operating a barge plus the losses of possible earned income) also impact the schedule of their next calls, and thus, the shippers' costs [42]. Is the total amount of these costs enough to change priorities in the allocation of handling resources between a barge and a ship for a stevedoring company? Surely not, due to the vessels' size, the volume of containers to operate and the frequency of transactions between owners and handling operators. But it should be enough to incite handling companies to propose solutions to barge operators, for the mutual advantage of ship-owners and shippers.

In some cases (e.g., for demurrage), the externality is known and "internalized" in the contract. The charter voyage warrants an over payment for barge operators as soon as their barges have to stay longer than it is necessary due to unusual long handling operation time (art. 2.13 of the French decree 96-855). But it is clear that this externality is underestimated in chains under tension like containers'. Any delay impacts the entire chain, requiring usually a buffer to cushion any hazard. Thus, if the voyage charter pays any inadequate delays (fixed over two or three days and a half by contract, according to French art. 9.1 of the decree 96-855) with demurrage, this adequate delay far exceeds those needed by containers (a few hours). Hence, delays are not correctly integrated in CBT.

Even if these delays are out dated, a second issue arises. The demurrage can't be paid to barge operators, since shippers do not operate handling with their own equipment and workers. Shippers ask barge operators to turn towards those responsible for delay: the handling operators who have no contract with CBT! A vertical integration by/of a stevedoring company could be an answer to the need for a centralized coordination between agents, as discussed in many papers about Rotterdam or Antwerp [7, 12, 14, 49].

In the French context of CBT development, these unsolved problems are major obstacles. The new Extended Gate Model, such as developed by the maritime stevedoring company ECT in Rotterdam and along the Rhine river, could be an adequate answer to coordination problems in barge transport in France.

References

1. Beelen M (2011) Structuring and modelling decision making in the inland navigation sector. PhD Dissertation, University of Antwerp
2. Bernadet M (2007) Evaluation socio-économique du projet de canal Seine-Nord Europe. Transports 442:87–97
3. Blauwens G, De Baere P, Van de Voorde E (2002) Transport economics, 2nd edn. De Boeck, Antwerp
4. Blum R (2010) La desserte ferroviaire et fluviale des grands ports maritimes français. Rapport à Mr le Premier Ministre, Paris
5. Bréhier O, Gavaud O, Guilbault M (2009) Les chaînes organisationnelles dans le transport: Enseignements de l'enquête ECHO. Rapport du CETE de l'Ouest, Nantes
6. BVB (2009) L'avenir du transport de marchandises et de la navigation fluviale en Europe 2010-2011. Bureau Voorlichting Binnenvaart, Rotterdam
7. Caris A, Macharis C, Janssens GK (2011) Network analysis of container barge transport in the port of Antwerp by means of simulation. J Transp Geogr 19:125–133
8. Cour des Comptes (2006) Rapport public thématique sur les ports français face aux mutations maritimes : l'urgence de l'action. http://www.ccomptes.fr/fr/CC/documents/RPT/RapportPorts Francais.pdf. Accessed 31 Jan 2012
9. CGEDD (2010) Evolution du fret terrestre à l'horizon de 10 ans. Rapport n° 007407-01 du Conseil Général de l'Environnement et du Développement Durable, MEEDDM, Paris
10. De Langen PW, Van der Horst MR, Koning R (2006) Cooperation and coordination in container barge transport. In: Puig J, Marcet i Barbé R, Carcellé VG (eds) Maritime Transport, vol 3, Technical University of Catalonia, Museu Maritim, Barcelona, pp 91–107
11. De Langen PW (2007) Stakeholders, conflicting interests and governance in port clusters. Res Transp Econ 17:457–477
12. De Langen PW, Douma A (2010) Challenges for using ICT to improve coordination in hinterland chains. In: International transport forum, round table on information and communications technologies for innovative global freight transport systems, Genoa, Italy, 8–9 Mar 2010
13. Demsetz H (1967) Towards a theory of property rights. Am Econ Rev 57(2):347–359
14. Douma A, Schutten M, Schuur P (2009) Waiting profiles: an efficient protocol for enabling distributed planning of container barge rotations along terminals in the port of Rotterdam. Transp Res Part C 17:133–148
15. Fernandez A, Arrunada B, Gonzalez M (1998) Contractual and regulatory explanations of quasi-integration in the trucking industry. Paper presented at the 14th annual conference of the European Association of Law and Economics, p 42
16. Finon D, Perez Y (2007) The social efficiency of instruments of promotion of renewable energies: A transaction-cost perspective. Ecol Econ 62(1):77–92
17. Fischer J, Monadier P, Allais V (2003) Amélioration des conditions d'accès et de trafics fluviaux dans les ports et zones maritimes. Rapport n° 2003-0004-01 du Conseil Général des Ponts et Chaussées, Paris
18. Fischman M, Lendjel E (2011) Efficience du marché et contrats types: une analyse transactionnelle du contrat type d'affrètement au voyage dans le transport fluvial de fret. Les Cahiers Scientifiques du Transport 60:7–38
19. Franc P, Van der Horst MR (2010) Understanding hinterland service integration by shipping lines and terminal operators: a theoretical and empirical analysis. J Transp Geogr 18:557–566
20. Frémont A (2009) Le développement du transport combiné pour les conteneurs: les enjeux. In: Proceedings of the Journées Réseaux Sécurité Transports, MEEDDAT, Paris, 2 June 2009
21. Frémont A, Franc P (2008) Voies navigables et desserte portuaire. Rapport final PREDIT, Paris
22. Frémont A, Franc P, Slack B (2009) Inland barge services and container transport: the case of the ports of Le Havre and Marseille in the European context. Cybergeo Eur j Geogr 437:1–20

23. Fremont A (2008) Les transports en France. Quelles mobilités pour quelle société? (La Documentation photographique)
24. Fu Q, Liu L, Xu Z (2010) Port resources rationalization for better container barge services in Hong Kong. Marit Policy Manag 37(6):543–561
25. Galbrun X, Le Du E (2007) 100 ans d'Union au service des ports français: 1907–2007. UNIM, Paris
26. Grégoire R (1983) Schéma de développement du transport fluvial et schéma directeur des voies navigables; rapport de la commission présidée par Mr Grégoire. Ministère des Transports, Ministère du Plan et de l'Aménagement du Territoire, Paris
27. Gouvernal E, Lotter F (2001) L'offre de services portuaires; Evolution des systèmes institutionnels et nouvelles formes d'organisation. Proceedings of Eficacia Logistica Portuaria, University of Curitiva, Brésil, Oct 2001
28. Guilbault M (ed) (2008) Enquête ECHO «Envois-CHargeurs-Opérateurs de Transport»: résultats de référence. Synthèse INRETS 56, Arcueil
29. Hislaire L (1994) Dockers, corporatisme et changement. Compagnie Française de Presse, Paris
30. IAU (2008) La place de l'Île-de-France dans l'Hinterland du Havre : le maillon fluvial. Institut d'Aménagement et d'Urbanisme en Ile-de-France, Paris
31. ITF (2009) Integration and competition between transport and logistics business. International Transport Forum, Discussion paper n° 2008–20098
32. Joskow P (1987) Contract duration and relationship-specific investments: empirical evidence from Coal markets. Am Econ Rev 77(1):168–185
33. Klein B (1988) Vertical integration as organizational ownership: the fisher body-general motors relationship revisited. J Law Econ Organ 4(1):199–213
34. Konings R (2007) Opportunities to improve container barge handling in the port of Rotterdam from a transport network perspective. J Transp Geogr 15:443–454
35. Konings R, Priemus H (2008) Terminals and the competitiveness of container barge transport. J Transp Res Board 2062:39–49
36. Masten SE, Saussier S (2002) Econometrics of contracts: an assessment of developments in the empirical literature on contracting. In: Brousseau E, Glachant J-M (eds) The economics of contract: theories and application. Cambridge University Press, Cambridge, pp 273–290
37. Merckx F, Notteboom T, Wilkelmans W (2004) Market-oriented information systems in inland barge transport. In: Sun L, Notteboom T (eds) Proceedings of the first international conference on logistics strategy for ports, Dalian, China, 22–26 Sept 2004, p 288–299
38. Minvielle E (2007) Transport fluvial de marchandises en France, un contexte favorable à la croissance. Notes de Synthèse du SESP 165, p 8
39. Monteverde K, Teece DJ (1982) Appropriable Rents and Quasi-Vertical Integration. J Law Econ 15(2):321–328
40. Notteboom T (2002) Consolidation and contestability in the European container handling industry. Marit policy Manage 29(3):257–269
41. Notteboom T (2004) Network dynamics in container transport by barge. Belgeo 4:461–477
42. Notteboom T (2006) The time factor in liner shipping services. Marit Econ Logistics 8:19–39
43. Notteboom T, Rodrigue J-P (2009) The terminalization of supply chains: reassessing the role of terminals in port/hinterland logistical relationships. Marit policy and Manage 36 (2):165–183
44. Panayides PM (2002) Economic organization of intermodal transport. Transp Rev 22 (4):401–414
45. Revet C (2011) Rapport d'information sur la réforme portuaire. Rapport n° 728 du Groupe de travail sur la réforme portuaire, présenté au Sénat le 6 juillet 2011, Paris
46. Rodriguez-Alvarez A, Tovar B, Wall A (2011) The effect of demand uncertainty on port terminal costs. J Transp Econ Policy 45(2):303–328
47. Soe S (2012) Memento de statistiques des transports. Paris: SOeS, Ministère de l'Ecologie, de l'Energie, du Développement durable et de la Mer. http://www.statistiques.developpement-durable.gouv.fr/donnees-densemble/1869/873/memento-statistiques-transports.html. Accessed 16 Jan 2013

48. Turnbull P (2006) The war on Europe's waterfront—repertoires of power in the port transport industry. Br J Ind Relat 44(2):305–326
49. Van der Horst MR, De Langen PW (2008) Coordination in hinterland transport chains: a major challenge for the seaport community. Marit Econ Logistics 10:108–129
50. VNF (2009). Guide du conteneur fluvial. Paris: Voies Navigables de France, www.vnf.fr. Accessed 16 Jan 2013
51. Williamson O (1985) The economic institutions of capitalism. The Free Press, New York
52. Williamson O (1996) The mechanisms of governance. Oxford University Press, Oxford
53. Williamson O (2010) Transaction cost economics: the natural progression. Am Econ Rev 100 (3):673–690
54. Zhao W, Goodchild AV (2010) The impact of truck arrival information on container terminal rehandling. Transp Res Part E 46:327–343
55. Zurbach V (2005) Transports de conteneurs sur le Rhin: quelles logiques de fonctionnement? MSc dissertation paper, Paris XII-ENPC-INRETS, Arcueil
56. Zurbach V (2006) Logiques d'acteurs et transport fluvial de conteneurs sur le Rhin. Note de Synthèse de l'ISEMAR 83:1–4

Appendix

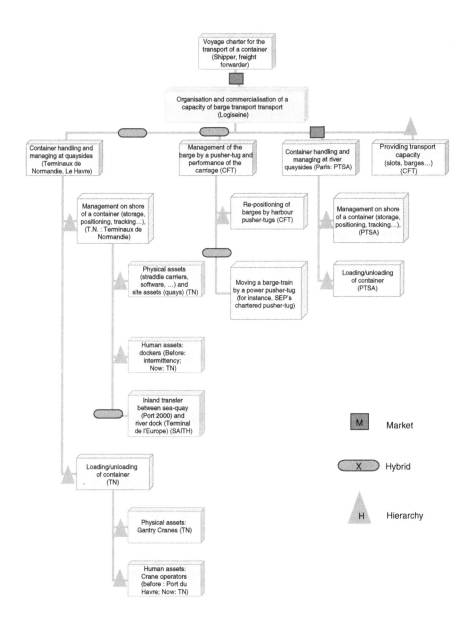

© The Author(s) 2014

89

A. Hyard, *Non-technological Innovations for Sustainable Transport*,
SpringerBriefs in Applied Sciences and Technology, DOI 10.1007/978-3-319-09791-6